The Most Defiant Devil

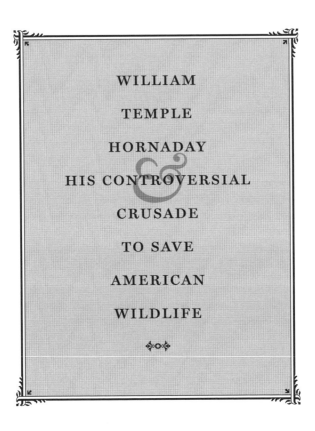

The Most Defiant Devil

Gregory J. Dehler

UNIVERSITY OF VIRGINIA PRESS   CHARLOTTESVILLE AND LONDON

*Lovingly dedicated
to the memory of my parents*

University of Virginia Press
© 2013 by the Rector and Visitors of the University of Virginia
All rights reserved
Printed in the United States of America on acid-free paper

First published 2013

1 3 5 7 9 8 6 4 2

LIBRARY OF CONGRESS CATALOGING-IN-PUBLICATION DATA

Dehler, Gregory J., 1968–
 "The most defiant devil" : William Temple Hornaday and his controversial crusade to save American wildlife / Gregory J. Dehler.
   pages   cm
 Includes bibliographical references and index.
 ISBN 978-0-8139-3410-5 (cloth : alk. paper)—ISBN 978-0-8139-3434-1 (e-book)
 1. Hornaday, William T. (William Temple), 1854–1937. 2. Naturalists—United States—Biography. 3. Zoologists—United States—Biography. 4. Wildlife conservationists—United States—Biography. I. Title.
 QL31.H67D44 2013
 590.92—dc23
 [B]
                                                                    2012046590

Title page illustration: William Temple Hornaday around the time of his return from his collecting expedition to Asia in 1879. *Courtesy of the Guilford Township Historical Collection, Plainfield-Guilford Township Public Library.*

CONTENTS

Acknowledgments	vii
Introduction	1
1. Iowa Farm Boy	9
2. Collecting Naturalist and Hunter	31
3. Stuffed and Living Animals	52
4. Director of the Bronx Zoo	74
5. Campfires and Conservation	95
6. Our Vanishing Wildlife	116
7. The Great War	137
8. Fighting the Establishment	158
9. Fighting to the End	179
Epilogue	199
Notes	203
Bibliography	229
Index	245

Illustrations follow page 94

ACKNOWLEDGMENTS

This book started nearly twenty years ago when I walked into Dr. Richard Harmond's office at St. John's University and asked him for advice on a master's thesis topic covering the Gilded Age and Progressive Era. He rattled off three names: Speaker of the House Joseph Cannon, the author Thornton Burgess, and the conservationist William Temple Hornaday. I selected Hornaday for the most pragmatic reason: The Bronx Zoo, which was a short drive from my home on Long Island, possessed the bulk of his professional papers.

A project this long incurs many debts, and I would be remiss if I did not thank all those who assisted me in one form or another. I wish to thank Dr. Harmond for placing me, however unwittingly, on the path of a truly captivating, if at times maddening, subject; Steven Johnson, former librarian at the Wildlife Conservation Society, who provided great help with Hornaday's enormous bulk of material; Tom Wells, Kurkpatrick Dorsey, Susan Murray, and the anonymous reviewers for their invaluable suggestions that directly shaped the final product; Christie Lowrance for our discussions of Hornaday and Burgess; Steven Hornaday for helping with his (and William Temple's) interesting family history; the late Peter Edge for taking the time to talk to me in 1999 and for sharing his mother's unpublished autobiography; Stephen Haynes of Minneapolis, Minnesota, executor, Estate of Emily J. (Bates) Haynes and owner of the papers of Chester E. Jackson, who contacted me out of the blue and shared his great-grandfather's valuable collection; Boyd Zenner and the staff of the University of Virginia Press; and all the librarians and archivists who over the years answered queries, sent photocopies, and offered suggestions. I especially want to thank Dr. Stephen Cutcliffe of Lehigh University. He is a patient editor, a good friend, and a model mentor.

Finally, I could not have completed this book without the encouragement and assistance of my family. I am truly grateful to have had parents who valued education and supported me in every way possible. My wife and our two children have demonstrated remarkable forbearance and patience as I buried myself in my office all too often to write the life of William Temple Hornaday.

INTRODUCTION

I repeat, and urge you to remember the fact that I am not a repentant sinner in regard to my previous career as a killer and preserver of wild animals, but I am positively the most defiant devil that ever came to town. I am ready and anxious to match records for my whole 76 years with any sportsman who wishes to back his record against mine for square dealing with wild animals.

WILLIAM TEMPLE HORNADAY TO ROSALIE EDGE,
11 NOVEMBER 1931

THE YEARS 1800 to 1900 were a bloody century for American wildlife. Prolific species like the ubiquitous passenger pigeon and the hardy buffalo were gunned down with reckless abandon. A determined group of conservationists managed to pull the buffalo back from the brink of extinction, but the passenger pigeon fell into the abyss. In 1914, Martha, a passenger pigeon who was the lone survivor of a species that may have numbered as many as 5 billion when Columbus landed, died in a Cincinnati zoo.

Various local extinctions mirrored the national exterminations of the buffalo and passenger pigeon. Throughout the country, antelope, deer, elk, turkey, and quail were just some of the species completely driven from the townships, counties, and states they had once inhabited.

In a burgeoning national economy, the railroad supplied wildlife products to the rapidly growing urban marketplaces. The buffalo provided warm robes and meat for those living in Chicago, New York, Boston, Philadelphia, and other metropolitan areas far from the Great Plains. Egrets and herons were mercilessly shot for their plumage, which was used to adorn the fancy hats popular with society ladies in the late nineteenth and early twentieth centuries. Market hunters shot massive numbers of ducks, geese, and a variety of upland birds like quail and grouse, and shipped them via railroad to men like Brooklyn's August Silz, who sold the meat throughout New York City. Even the seals in the faraway Pribiloff Islands off the coast of Alaska faced an existential threat when Canadian and Japanese sealers took to the sea and killed the seals on water (a wasteful practice known as pelagic sealing).[1]

This slaughter did not proceed without protest, however. Increasingly in the late nineteenth century, people voiced opposition to the killing of American wildlife. Among those speaking out were gentleman sportsmen hunters who decried market hunting; scientists who disparaged the wasteful use of resources and sought to rationalize the killing; Audubon Society women who valued songbirds; and hikers and mountain climbers who formed organizations like the Sierra Club in California to protect their wilderness retreats. The protesters also included urbanites who desired to escape their overstimulating environments, bird-watchers devoted to species of no economic value, and thousands of concerned citizens who voraciously consumed the books of such authors as John Burroughs, Jack London, John Muir, Ernest Thompson Seton, and Mabel Osgood Wright in the comfort of their urban parlors.

This eclectic group was united by a common perception that nature was a collective commodity. Their outlook was a clear expression of the Progressive Era ethos of placing the greater good of society before the gains of a few individuals and businesses. Like the larger Progressive reform effort, the conservation movement placed most of the blame for the ills of the land squarely on greed. As the larger Progressive Era reform movement sought to level the playing field through the application of state power to regulate incomes, labor practices, social habits, and the marketplace, conservationists sought to employ the same means to protect the environment. In both cases, they relied on the opinions of established experts to shape policies.[2]

The Progressive conservation movement racked up impressive accomplishments between 1900 and 1920. Its various interests, constituencies, and allies exponentially expanded the forest reserves and placed them under the expert management of the Forest Service, added national parks, enlarged the scope of protection under the Antiquities Act, established numerous bird and wildlife sanctuaries and refuges, and put hunting laws on the books. They also returned the buffalo to the prairie, created protections for all migratory birds, brokered treaties to conserve resources across international boundaries, and developed a national conservation policy. In the process, conservationists dramatically increased the role of the federal government in areas traditionally reserved for the states.

But despite the achievements of the Progressive conservation movement, serious tensions developed among its various constituencies. Increasingly, this diverse conglomeration of interests found itself divided over goals, policies, and even values. Sometimes the disputes became vicious, bitter public squabbles. Between 1908 and 1913, for example, a battle raged between those who wanted to dam the Hetch-Hetchy Valley in Yosemite National Park to provide water for the city of San Francisco, and those who saw such damming as despoliation of nature. Prominent conservationists stood on both sides of the argument. The dispute made its way into newspaper headlines and the deliberations of congressional committees. Chief Forester Gifford Pinchot supported the dam as a wise use of nature to benefit humanity, while John Muir of the Sierra Club and Robert Underwood Johnson of *Century Magazine* bitterly opposed it on the grounds that national parks were inviolate sanctuaries.[3]

A similar drama played out on a smaller scale over protection of the fur seal of Alaska between 1909 and 1913. Scientists of the Department of Commerce and Labor's Fur Seal Advisory Board, and protectionists led by William Temple Hornaday and the Camp Fire Club fought a battle royal that became personal. While both sides agreed on the necessity of stopping pelagic sealing through an international convention, they clashed sharply over the best policy for regulating land sealing.

By this time, the days of the autodidactic general naturalist wandering the woods with a gun and notebook were over. Specialized fields like anatomy, biology, and zoology had superseded the old-fashioned catchall known as natural history. Sportsmen, too, had changed. By 1913, states had adopted most of their proposed restrictions on hunting, and the market gunner was largely driven out of business. Many leading sportsmen associations became concerned that the pendulum was swinging too far in the other direction, toward overregulation. These sportsmen associations clashed with the growing number of nature lovers working to restrict hunting, whom the sportsmen regarded as "sentimentalists."

Meanwhile, the number of nature lovers who neither hunted nor pursued specialized scientific interests grew with the rise of the consumer society and urbanization. They included the tens of millions of Americans who went to zoos on Sunday, bought their food at a supermarket chain during the week, and drove their Model T on camping trips on vacation. They developed a softer approach to wildlife that at times conflicted with the efforts of the sportsmen and scientists within the conservation movement.

This biography examines the life of one of the key players in the Progressive wildlife-protection movement, William Temple Hornaday. Hornaday was a prominent spokesman for the millions of Americans who had grown to appreciate the aesthetic value of nature and become concerned that nature was being threatened by crass economic calculations. As the first director of the Bronx Zoo (he deplored this vulgar name for the institution, preferring the more distinguished New York Zoological Park) from 1896 until 1926, he shaped the views of millions of Americans who visited his zoo each year. His impact on zoo design was considerable. Under his influence, zoos came to embrace education, science, and conservation as core elements of their missions. The Bronx Zoo set an example that others followed, and Hornaday pressured his adherents to conform to his vision. Yet, despite his significant influence and his successes as a zoo director, Hornaday's vision was often suspect and controversial, and his personal shortcomings were striking. He waged a losing battle to prevent the adoption of Carl Hagenbeck's pioneering zoo reforms in the United States, and frequently exhibited a showmanship that crossed the boundaries of good taste. He could be difficult to deal with.

During his lifetime, Hornaday was credited with restoring the buffalo to its former habitat in the Great Plains. The near-extermination of the buffalo in the late nineteenth century shaped Hornaday's conservation ethic in several important ways. He was shocked and outraged that such a strong and virile beast numbering in the tens of millions could be hunted out of existence in two decades. If that could happen to such a mighty species, he reasoned, surely weaker ones were almost guaranteed to become extinct if they did not receive adequate, if not radical, protection. The fate of the buffalo convinced Hornaday that hunting for profit was the prime cause of wildlife decline. As firearms technology improved, new modes of transportation conquered nature, and growing cities offered a hungry marketplace for meat, humanity's capacity to kill came to vastly outpace any species' ability to reproduce. This Malthusian equation of wildlife population and the lessons of his Adventist upbringing governed Hornaday's conservation philosophy.

As a prolific writer, Hornaday authored over a dozen books, scores of articles and editorials, and thousands of letters on wildlife conservation. He waged his campaigns in much the same way as the Adventist temperance movement championed Prohibition. Like the Women's Christian Temperance Union, he exhorted the public to do the right thing, hoping individuals would find the wisdom to join the good fight. If this failed, however, he had no compunction about mandating correct behavior through the laws of the state. Like the temperance movement, he believed selfish conspiracies of moneyed interests thwarted progress.

Through his writings and other activities, William Temple Hornaday helped define the dichotomy within the Progressive Era conservation movement between those who, like Chief Forester Gifford Pinchot, sought to use nature wisely and efficiently for human ends, and those like John Muir who valued the preservation of nature over its use. Historians usually refer to the former perspective as utilitarian and the latter as preservationist. Hornaday helped sharpen the preservationist ethic for wildlife in much the same way that Muir did for the larger conservation movement. While there are enormous differences between the two men, Hornaday and Muir both made the case that nature had a right to exist and prosper without human interference, and that humanity would draw greater long-term benefits from an unspoiled nature than it would gain from a more immediate exploitation of natural resources.

But Hornaday fought his enemies more bitterly and more often than Muir, making him a much more divisive figure. "My policy is no quarter to the destroyers of wildlife," he wrote in 1913. Increasingly, Hornaday attacked other conservationists, especially the utilitarian sportsmen, as the true destroyers of wildlife, fighting them with greater zeal than he did those who eschewed any form of natural-resources conservation. His ethos for preservation was rooted firmly in what he felt was humanity's moral responsibility to protect wildlife for future generations. He thereby built a bridge between the Progressive Era conservation movement and the post–World War II environmental movement. Such later laws as the Sea Mammal Protection Act (1972), Endangered Species Act (1973), and Alaska Conservation Act (1980) emphasized moral responsibility in a way that Hornaday would have appreciated.[4]

Hornaday considered himself the public's watchdog over scientists on a government payroll, a group he tended to viscerally distrust. He became the bane of cabinet secretaries and government bureau chiefs, challenging both their motives and their policies. Hornaday publicly aired his accusations of government misconduct in newspapers and congressional testimony. He attacked numerous state boards and game commissions, as well as such federal agencies as the Fur Seal Advisory Board, the Migratory Bird Advisory Board, and, most importantly, the U.S. Bureau of Biological Survey, the forerunner of the U.S. Fish and Wildlife Service. He criticized these agencies for lack of imagination, bureaucratic malaise, and naively falling under the evil influence of murderous profiteers. They were, in his mind, complicit in the destruction of America's wildlife.

In challenging scientific authority, Hornaday broke with his contemporaries who valued expert advice. He seemed out of touch to them, and even threatening, when in the 1920s he seriously proposed taking wildlife regulation from the Department of Agriculture and giving it to U.S. congressmen, who were untrained in such matters. As with his moral argument for protecting wildlife, his distrust of government scientists who were uninfluenced by public opinion was a harbinger of the post–World War II environmental movement, especially Rachel Carson, who famously challenged pesticide spraying in *Silent Spring* (1962). His distrust also presaged the attitudes of those who challenged conventional wisdom on such matters as atomic fallout, nuclear energy, chemical waste, and carbon emissions that the government first told the public presented little or no threat to their wellbeing.

Hornaday questioned the growing consumer economy and its effect on nature at a time when many praised the New Era of the 1920s for its booming stock market, expansion of consumer choices, and new laborsaving inventions. As early as 1888, he had decried the growth of technological innovations that allowed mankind to kill wildlife much more quickly than it could breed. He lumped guns, camping outfits, scouts, arms manufacturers, advertisers, tourism agents, railroads, and automobiles in the same category as the market hunter and game hog. Instead of seeing the automobile as a sign of independence and freedom, he considered it a means to make hitherto safe wildlife habitats and breeding grounds more accessible to better-armed and -equipped hunters. Other conservationists looked at consumption with a suspicious eye, as the efforts to restrict the sale of game meat and ban plumage attest. However, few conservationists criticized consumption as much as Hornaday or, indeed, made it the center of their ideology. Furthermore, Hornaday never blithely dismissed the demise of the buffalo or any other species as simply a consequence of the "advance of civilization" as did some other conservationists. Those concerned today about the effect of consumerism on the warming of the globe, the effects of chemicals on our health, and our ability to safely dispose of nonbiodegradable waste can regard Hornaday as their intellectual grandfather.

Like the Progressive Era reform movement itself, Hornaday and other conservationists were blinded by racial bias. From Madison Grant's "scientific" racism to Theodore Roosevelt's fear of race suicide, conservationists by and large saw a parallel between a nature under siege by base financial interests and a culture at risk of being overwhelmed by undesirable immigrants from southern and eastern Europe. Conservationists who urged protection of wildlife native to America also proposed extermination of invasive species, such as sparrows and starlings. And they lobbied states to adopt alien gun restrictions in an effort to keep immigrants, whom they considered likely "game hogs," out of the field. For many of these strong-willed, well-bred, patrician elites, saving the buffalo represented the preservation of American culture.

Hornaday's racial views differed from both Grant's and Roosevelt's in that Hornaday did not subscribe to the theory that genetics proved one race superior to another. Nor did he consider the Anglo-Saxon race existentially threatened by other races. Hornaday believed cultural environments shaped people. After nearly a year in India, he departed convinced that the people were lazy and shiftless. But he was equally convinced that the civilizing project of the British Empire, as he saw it, could reform the culture of its subject people.

Hornaday's attitudes toward ethnicity and race affected his wildlife conservation. He attributed an unnatural desire to harm the most delicate and beautiful species to those races he looked down on the most. He blamed blacks for killing almost all the songbirds of the South, and he used this association to advance the cause of wildlife conservation. In his most famous work, *Our Vanishing Wild Life* (1913), he included photographs of African American hunters in a cynical effort to overcome southern opposition to a sizeable extension of federal power over wildlife protection. He expected these photographs would resonate with white southern legislators. He thought Italians equally dangerous to American wildlife as blacks because they, too, killed the songsters, especially after he caught several Italian immigrants shooting birds on the grounds of his zoo.

The Native Americans, on the other hand, seemed to Hornaday simply beyond redemption. Unlike other conservationists, he did not consider Native Americans as proto-ecologists who killed only what they needed and used all that they killed in the most efficient manner for their ecosystem. Instead, he viewed them as "game hogs" as well, ones that persisted in killing the declining buffalo despite the U.S. government's benevolent food subsidies to them.

1

IOWA FARM BOY

WILLIAM TEMPLE HORNADAY was born into a family of mythmakers. The origins of the Hornaday clan have baffled generations of family genealogists who have been unable to reliably trace the family name to Europe. Various family legends claimed that the Hornadays immigrated to America from England, Northern Ireland, Germany, Norway, Hungary, and even Spain, but none of these stories can be fully substantiated. William Temple Hornaday himself believed that his ancestors were Englishmen who immigrated to Northern Ireland after Oliver Cromwell's fall, before setting sail for America. To account for their untraceable past, several origin myths emerged. In one such story, two wealthy brothers escaped the heavy burdens of their aristocratic birth by sailing to the colonies and creating an entirely new name for themselves. In another tale, an orphaned cabin boy who dispensed water once a day in a hollowed-out horn to the ship's passengers and crew earned the nickname "horn-a-day," and adopted it as his own surname. Certainly, the truth of the family origins must lie somewhere between aristocrats and poor orphans, but no one knows for sure.[1]

The first Hornaday in America, John, appears on a tax table in Orange County, North Carolina, in the early 1750s. This location suggests that the Hornadays, like the great bulk of their neighbors, emigrated from the borderlands of England, although of course it does not explain the uniqueness of the name itself. The borderlanders adhered to dissenting Protestant faiths and continuously demonstrated a spirit of rebelliousness against all forms of centralized authority. Individual convictions strongly favored personal independence through militant means. It was natural for the borderlanders to seek refuge on the frontier of the American colonies, where they made lives for themselves as far from the king's tax collectors and clergymen as they could. A common man now visible to us only in outline, John seems to have been as rebellious as his neighbors. He was a frontiersman who flirted with the Regulator movement in the late 1760s and owned a few slaves. At some point during the American Revolution, he became estranged from his wife, Christian. He died sometime between the censuses of 1800 and 1810 while farming in South Carolina.[2]

John's children dispersed throughout the Carolinas, but his son Nathan sold his slaves and immigrated north to Ohio in his middle age at the insistence of his second wife, a Quaker. Nathan's son Ezekiel, born in 1795, took the Hornaday name farther west, settling in central Indiana in the early 1820s. Ezekiel embraced the great wave of religious revivals known as the Second Great Awakening that swept through the country after the War of 1812, and helped found the Christian Church of Plainfield, Indiana, in 1830. A moderately successful farmer by the standards of the day, Ezekiel amassed more than 1,000 acres of prime farmland in the White Lick Valley before he died in 1876.[3]

Born in 1818, William, the eldest of Ezekiel's twelve children, did not share his father's good fortunes in health or land. William's wife, Orpha Hadley, died in 1851, leaving him with a daughter, Mary, and three sons, Calvin, Clark, and Silas. The following year he married Martha Varner Miller, a widow, with two sons of her own, David and Minos. "He and my mother were a fine and noble pair," their youngest child recalled. They added a daughter, Margaret, in 1853, and a son, William Temple, who was born on December 1, 1854 in the walnut-paneled, mahogany-furnished home of his paternal grandfather.[4]

The Varner family had a less mythic origination story, but the Pennsylvania Dutch blood of the Varners and the Scotch blood of the Pughs, William Temple Hornaday's maternal grandmother's family, flowed through William Temple's veins as well. He seems to have had a closer relationship with his mother's parents than his father's. Although Ezekiel survived his son, William Temple wrote little of him, while he characterized "Grandmother Varner" as "a wise woman, of strong and sterling character, and truly fit to help found a pioneer state."[5]

Although William Temple Hornaday described his father as a progressive farmer who kept abreast of the latest agricultural advances and frequently adopted innovative techniques, William Sr.'s Hendricks County land was located in the worst agricultural area in the state. Its clay soil proved largely unproductive, regardless of the methods employed to improve its yield. In 1857, William Sr. decided to do what his father and grandfather had done before: go west in search of cheap lands. He took the Hornaday name to Iowa.[6]

The wagon train, which consisted of several branches of the Hornaday family, lumbered slowly through Indiana and Illinois before crossing the Mississippi River into Iowa. At one of the railroad crossings, Billy, as William Temple was called in his youth, got scared by the shrieking noises of an "iron monster rushing straight for us" and hid under the team of horses pulling the wagon. As the train approached, the horses grew restless, then frantic. Martha's excited screams caught the attention of one of the men in the company, who pulled Billy out before the horses trampled him. He claimed that he could still remember his mother's screams almost eighty years later. It might have been the first time William Temple Hornaday cheated death, but it would certainly not to be the last.[7]

After crossing the Mississippi River into Iowa, William Sr. settled his family into the southeastern part of the state and purchased a 270-acre plot in Eddyville, Mahaska County, not far from Miller Creek. Iowa was not the land of milk and honey for William and Martha as they struggled against nature, ill-health, and loneliness in their new home. The winter of 1859 was especially difficult. Tragedy struck when Billy's fourteen-year-old half brother, Silas, died in November. A harsh and snowy winter followed, and many families struggled to survive. "Some of our neighbors are next thing to suffering," Martha wrote to her parents. "But we can't help them much for they are too numerous and the time until harvest too long to risk much in the way of charity for we remember charity begins at home." Drought and torrential downpours defined the following year of 1860. Billy either chose not to share these pre–Civil War experiences, which clashed so sharply with what he described in his unpublished autobiography, "Eighty Fascinating Years," as "halcyon days," or he simply had no memory of them. Yet, his own stubbornness and independence made matters worse as he drove his mother to the point of despair. Billy "is an awfull bad boy," Martha wrote in exasperation to her son David. He "knows how to hold his own in a quarrel as well as common boys does [*sic*] at 10 or 12 years old and no wonder for he has a great sum of practice." William Temple's streak of stubbornness was a defining characteristic of his personality, and it would prove just as exasperating to his allies, employers, enemies, and friends alike over the next seventy years as it had to his mother.[8]

William Sr.'s poor health intensified the family's struggles in Iowa. He suffered a constant and at times acute pain in his left breast, the result of an old injury that often prevented him from working his farm. Within two years of staking his claim in Iowa, William Sr. came to realize that the pain would not go away. As his wife noted in a March 1859 letter to her parents, "he never expects to be well like he once was." When William took a short, slow ride on horseback, it nearly crippled him. "The jar hurt his side so much that his whole system was put out of order for a week or more," Martha wrote to her mother one August. In the face of such pain, William relied on his older sons and a number of hired hands to do the hard physical labor while he managed the business side of the farm. This plan might have worked had war not engulfed the country.[9]

Like all families in the United States at the time, the Hornadays could not escape the effects of the American Civil War. Patriotism overwhelmed both William Sr.'s hired hands and his sons, and they flocked to the colors. William Sr.'s eldest son from his first marriage, Clark, enlisted with friends from Plainfield in the Nineteenth Indiana Regiment. Minos Miller, Martha's second son from her first marriage, who worked closely with William Sr. on the farm's business, also enlisted in the army. After a stint in an Iowa regiment, he became an officer in the Fifty-Fourth Colored Regiment. David Miller, Martha's eldest son, returned to Indiana to manage the family farm when Martha's brother Allen Varner enlisted. The depleted labor pool for William Sr.'s farm consisted of only Billy, who was too young for significant labor, and Calvin, William Sr.'s son from his prior marriage, who was thirteen in 1861. Calvin suffered from such painful bouts of rheumatism that his younger brother later described him as suffering "more than a thousand deaths."[10]

The war caused no end of worry among the adults in the Hornaday home. Martha fretted endlessly about the safety of her sons Minos and David, and followed Minos's troop movements in his letters and newspapers. After one battle in Arkansas in July 1863, however, she had neither received a letter from him nor seen the names of the dead. "As to poor Minos I fear he is gone forever," she forlornly told her parents. Minos survived the battle and the war, but she worried about him throughout. For David she concocted various schemes to help him evade the draft, including borrowing money to hire a substitute, but he never received the dreaded call to arms. Clark proved less lucky. Working with the wounded, he contracted dysentery and died in a Philadelphia hospital in early 1863, just weeks shy of his twentieth birthday.[11]

The spectacle of the war fascinated young Billy. He demanded souvenirs from Minos, who responded with a variety of trinkets, including Confederate bills and coins. Billy became cranky when he did not receive letters as frequently as he thought he should and admonished Minos for not writing often enough. But like most soldiers, Minos was hungry for news from home and just as frustrated with the slowness of the post system. By 1866, the whole issue of letters developed into a battle of wills between eleven-year-old Billy and Captain Miller. "I guess Billy has concluded he can remain silent as long as I can," Minos wrote in August 1866. "Tell him I admire his spunk but think he might write once in a while and tell me what he thinks about me not writing to him." The dispute clearly frustrated Martha, who like her husband was suffering from recurring physical ailments: she informed Minos that Billy had in fact written him recently, but that the letter must have gotten lost in the mail, as so many did during the war and its immediate aftermath. Indeed, several letters sent back and forth between Minos and his family in Iowa, including ones with money, never reached their intended destination.[12]

The misunderstanding between Billy and Minos over the letters was indicative of a pattern of behavior that started very early in Hornaday's life and continued throughout. He often became inflamed over relatively minor matters and took affront at things that he interpreted as personal slights. The triggers for the personal disputes that resulted could be circumstantial, unintended, and beyond the control of any one individual, such as lost letters or a soldier's lack of time to write. But when the other party failed to meet his rigid demands, and this occurred frequently, Hornaday's normally genial personality changed, and he became embittered and full of animosity.

The Hornaday family's financial struggles compounded Martha and William Sr.'s wartime worries over the safety of their children. Indeed, after half a dozen years of fruitless toil in Iowa, Martha wanted to return to Indiana. But William Sr.'s daughter Mary, his eldest child from his first marriage, convinced him to remain in Iowa. Instead of moving east, William Sr. attempted to turn his farm into a less labor-intensive cattle ranch. But he could not raise the capital required for the conversion. "William is trying to get money from Minos or anybody he can to buy cattle with but if Minos let him have any I hope he will do better than he did with his other money but I say nothing to nobody," Martha confided to her son David. By the end of the year, he sold the farm and purchased a twenty-acre plot in Knoxville, twenty miles to the west of Eddyville, in Monroe County.[13]

Soon after William Sr.'s decision to stay in Iowa, Martha's health completely broke down. A severe back injury was her most immediate problem. In May 1864, Billy wrote his half brother David to say, "Mother is getting better." But the boy failed to understand the gravity of the situation. "He seems to think I am better but if he knowed [sic] all he would not think so," Martha wrote to David after reading Billy's note. "He thinks because I can walk and sit up most of the time I am nearly well but I am far from well and if I am to believe the Dr. have no real hope of ever being well these five years to come." The doctor's inability to determine the source of Martha's pain and thus offer an accurate diagnosis compounded her sense of despair. William Sr. did not accept the situation at all. Though she advised him that "I could not do so much hard work and he might shape his affairs according," as she later related to David, "he would get mad and so would I."[14]

Martha and William Sr. turned to their faith for comfort. Hornaday hardly mentioned the subject of religion in "Eighty Fascinating Years," writing only in passing that the family's leanings were "profoundly moral; and significantly, but not painfully religious." Nonetheless, William Sr.'s and Martha's conversion to Seventh Day Adventism in 1863 had a powerful, if unconscious, effect on their son. They subscribed to the *Adventist Review* and other newsletters, invited preachers to meetings in their homes, and hosted days-long revivals on their farm. None of this escaped Billy's attention. To David, he described meetings held by a Reverend Snook in February 1864 as "fine meetings." These events, with their bonfires, charismatic preachers, picnics, and wagonloads of people coming to his house, left a deep and lasting impression on Billy. In 1926, he would tell U.S. Senator Samuel Shortridge of California how he remembered the legislator's father, a bearded Adventist preacher who gave "powerful" sermons. Reverend Shortridge frequently visited Martha and William's home.[15]

Although Hornaday drifted away from Adventism in adulthood and worshipped in other Protestant churches, three particular messages of Adventism left their imprints on him: absolutism, immediacy, and evangelism. Throughout his life, Hornaday understood people and events in absolute terms. Things were righteous or evil, black or white. There was no room for a middle ground with shades of gray, hence his inability to compromise on what he regarded as principles of morality. Many people who dealt with him, however, tended to view this behavior as inflexible and self-serving. Adventism's message of immediacy, as reflected in the belief in the imminent Second Coming of Jesus Christ and the need to atone for sins before it was too late, had an equally deep effect on Hornaday's view of the world. In later years, one can reasonably hypothesize, it led him to demand instant correctives through drastic measures, if required, because moral correction should never be postponed. Opponents frequently mistook this for simple impatience, but it was more than that. And all of Hornaday's educational work and conservation writings directed toward the general public had an evangelical flavor. One Iowan contemporary noted that an Adventist "carried a theological chip on his shoulder, waiting for discussion." That description fits Hornaday quite well. In later years, when he incorporated this evangelical style into his conservation writings, he was not merely laying out the facts before the reader. Rather, with pictures and grisly "truths," he was appealing to the emotions of his readers, and not simply seeking their support for a measure because it was the logical course, but proselytizing on the righteous cause of conservation. No matter the particular bill or issue, he was looking less for supporters than for lifetime converts. Frequently he framed his numerous appeals to the public as a preacher would have. He condemned bad and immoral behaviors, predicted dire consequences for failing to act, and raised the specter of fire and brimstone if no action were taken.[16]

Adventism had three more conscious impacts on Hornaday's thinking. First, the Adventism temperance reform movement provided a blueprint for his conservation campaigns. Both exhorted the public to do the right thing, praying the sinners would find enlightenment. If not, however, they had no qualms about enforcing correct behavior through the power of the state. Although he was not a strict teetotaler, Hornaday wrote a pamphlet in 1887 entitled *Free Rum on the Congo* for the Women's Christian Temperance Union. Second, Adventism was stridently anti-Catholic, an ugly disposition Hornaday maintained throughout his life. In 1876, he caused a ruckus in Northern Ireland and blamed it all on the backwards and superstitious nature he attributed to their faith. "They were *all* Catholics," he exclaimed to a friend, as if that explained everything. Finally, Adventists saw conspiracies working against their religious and social reforms. As temperance advocates believed there was a natural collusion of vested financial interests and Catholics working against their campaign against alcohol, Hornaday would later argue that gun makers, their conservation lackeys, and agents of the U.S. Bureau of Biological Survey actively subverted wildlife protection in a conspiracy to enhance their own profit.[17]

When not engaged in the backbreaking work of farming, the Hornaday males, like so many of their nineteenth-century contemporaries, liked to hunt and fish. Although hunting often provided food at the table, the real goal was camaraderie and outdoor recreation. It was a chance for men to get away from the labors of the land and enjoy time with each other away from the women. As William Jennings Bryan recollected of his own childhood in neighboring Nebraska, the fun started the night before, when father and son cleaned the guns, molded lead bullets, and exchanged stories from previous adventures. These rituals bound father and son closer together.[18]

As one might expect from this future conservationist and zoologist, Billy also developed values regarding wildlife. Several incidents implanted indelible memories. In one tale, he was visiting his uncle's house with his parents when a noisy blue jay buzzing about the cherry trees kept interrupting the conversation. The adults ordered Billy to drive the noisy bird away. He picked up a stone and threw it at the bird, striking it dead. He remembered thinking, "then it began to dawn upon me that the beautiful bird was DEAD, that I had taken its life, that I could not make it live again, that its poor dead body was of no use to anyone and would be wasted." Another tale recounted a squirrel hunt with his half brother Calvin and his father. After cornering a squirrel in a tree, Billy fired and killed it. A pang of guilt set in, according to his account. "It was unfair," he wrote, "and it was too easy to be real sport for me. Moreover, squirrel flesh had an odor that suggested rats, and its taste was not all good." He also recalled luring a prairie chicken into a trap. "I lost all interest in the business, and never set another trap. The baiting of cold and hungry birds with treacherous ears of corn was to me no sport; and we had an abundance of other meat for our table."[19]

The cumulative effect of these lessons was the belief that killing for the sake of sport was wrong, and that the taking of animal life had to serve some higher social purpose. Hornaday later killed thousands of animals on the grounds that well-stocked museums provided for the education of millions of people who could not see animals in the wild. When living animals in zoological parks replaced dead ones in museums, he greatly reduced his hunting trips, limiting them to those he deemed of scientific merit. Though one might question the necessity of these "scientific" trips, the need for a higher purpose was established early in his life.

Meanwhile, William Sr. and Martha encouraged their youngest child to read, with the result that he was literate before he entered a white one-room schoolhouse—a building no doubt resembling the one on the back of the Iowa state quarter—when he was six years old. The first book he read was *The Careless Chicken*. Proud that her five-year-old son had memorized most of the book, Martha wrote how they used it as a didactic tool in the home: "we call Bill the careless little chick he threatens to burn it but he likes to learn it the best of any book he has." Billy also read "Uncle Merry's" series of adventure stories. Martha cultivated an interest in the Bible, poetry, and satire as well. Billy remained inclined to these topics throughout his life, adding history and politics when he reached adulthood.[20]

Martha never recovered from her back injury. In 1866, she entered the Adventist Western Health Institute in Battle Creek, Michigan, as one of its first patients. During her convalescence, the institute was nothing more than a house with a capacity to accommodate fewer than ten patients. In due time, it would grow to become the world-famous Battle Creek Sanitarium. Neither the doctors nor the nurses at the institute could restore Martha's health, and she returned to Indiana, where she suffered debilitating bouts of coughing. Her health deteriorated to the point where she was unable to perform the simplest of tasks without assistance. "This morning I felt well enough to comb my hair myself before breakfast, which I have not often done in the last month," she dictated in a sad letter, quite possibly her last, shortly before her death on January 19, 1867. William Sr. struggled in Knoxville with his own declining health and sent Billy east to live with David Miller. On June 12, 1869, William Sr. died of what the coroner recorded as gastritis, an inflammation of the stomach lining. "By that final blow," Billy wrote of his father's death, "our family was left a shattered and shaken remnant."[21]

Cast out into the world as an orphan, Billy now became a family problem, and no one was quite sure what to do with him. In his will, William Sr. had made Ben Auten, a Knoxville neighbor and fellow Adventist, his son's financial and legal guardian. With his father's estate heavily encumbered with debt, Billy received no inheritance. Throughout his life, Hornaday worried constantly about personal finances, an obvious emotional scar from his overwhelming transformation into an orphan.

After William Sr.'s death, Billy circulated through several households in Iowa, Indiana, and Illinois. While in Knoxville, he lived with Minos and Ben Auten. In Indiana, he lived intermittently with David. At the time of the 1870 census, he was with Martha's brother Allen in eastern Illinois, where he was recorded as a farmhand. Captain Allen Varner, who served in the Twenty-Fifth Illinois Infantry, had a distinguished service record during the Civil War. His nephew fondly referred to him as the "Hero of Chickamauga," a bloody battle fought in Tennessee in the summer of 1863. Throughout his life, Civil War terminology colored Hornaday's descriptions, analogies, and prose.

Billy had one certain thought about his future: on no condition would he ever become a farmer. It is bitterly ironic that William Sr. broke his health ostensibly to leave land for his children, only for his two surviving sons to unequivocally reject a life of farming. While Billy followed his career course into taxidermy, zoology, and conservation, William Sr.'s other surviving son, Calvin, became a jeweler in Keokuk, Iowa. Ben Auten suggested dentistry to Billy. "Fortunately for us all, the plan failed to mature," Hornaday wrote. He considered a career in journalism and wrote to two local newspapers, but received uninviting responses. Ultimately, he decided to attend college before making a definitive career choice. He set his sights on the new University of Iowa at Ames.[22]

Having received no money from the estate of his father, Billy could not attend Ames unless he received a scholarship. Iowa allocated two scholarships per county, but those for his home county, Marion, had already been filled. A Knoxville friend of his father suggested to Billy that he attend nearby Oskaloosa College, his half brother Calvin's alma mater, where he could take needed preparatory courses and apply for one of the vacant scholarships from Mahaska County. He arrived at Oskaloosa for the fall term of 1870.

Oskaloosa turned out to be more than a way station to Ames. Billy received his introduction to taxidermy during his time there. He had already had a brief encounter with the subject. Once, when he and Calvin were hunting, they had shot a duck. They spent a good deal of time looking it over as a scientific specimen, and Billy later remembered that he wondered if there was a way to preserve it. Later, in an Indianapolis store with David, he got his answer when he spotted a stuffed duck on a shelf. But it was not until he sat in the Oskaloosa chapel for mandatory church service that summer that "a thunderbolt fell from the sky," as he later described the moment. The college president announced that the institution's museum required specimens and that any student who participated would receive taxidermy lessons as compensation. Billy immediately applied. To his disappointment, he received only one lesson, but it was enough to whet his appetite for more.[23]

In 1870, there was much to recommend taxidermy to an ambitious teenager. The combined effect of the Morrill Act, which created land grant colleges, and urban civic pride led to an explosion in museums. In the age before large zoological parks, museums often provided the only opportunity for people to see "wildlife." At the colleges, the specimens served as valuable scientific objects for students studying animals. As Billy reviewed his career options, taxidermy offered a field of growth, opportunity, advancement, and most of all, adventure. Many taxidermists collected their own subject matter, with a few even traveling to exotic locales in Asia and Africa.

In the spring of 1872, Billy arrived in Ames according to plan. Although the Iowa state legislature had created the college in 1858, the Civil War delayed completion of its first building for nine years. In 1872, the inaugural class of 173 entered a barren campus almost totally devoid of trees, vegetation, and shrubbery. Despite its spartan appearance, Ames proved to be a veritable garden for Billy's professional development. He truly found his way at college, thanks in large part to Charles E. Bessey, a twenty-seven-year-old Ohioan who served as chair of the Botany and Zoology Department. Hornaday described

Bessey as "my guide, philosopher, and friend" and his tutelage under him as "a leading event of my life."[24]

In his zoology classes, Bessey relied on Sanborn Tenney's recently published *Manual of Zoology* as a text. Tenney stressed the importance of whole animals instead of breaking them down into smaller internal systems. This suited Hornaday, who wanted to understand an animal's adaptability, diet, habitat, range, and relationships to other animals, not its cellular mechanics. Thirty years after he left Ames, Hornaday thanked Bessey for putting him on this path. "I have been very grateful for the fact that you started me exactly as you did," he wrote, "and did not discourage me by a long dose of anatomy, embryology, and gropings among the lower forms of life at a time when I wanted practical knowledge of beasts and birds and creeping things." In 1904, Hornaday contributed his own zoological text emphasizing "practical knowledge," *American Natural History*, a popular work that he revised and reprinted in several forms until his death in 1937. His view of nature encouraged manly and rugged study of animals in the wild, often with a rifle. It was an old view, going out of fashion even as Hornaday was learning it. But this view of zoology remained frozen in place for him despite all the learning with which he would come into contact over the ensuing decades. He deplored later scientists who attempted to move the study of zoology out of the wild and into the laboratory, and who forsook the study of whole animals to scrutinize their biological systems and cellular building blocks.[25]

Equally important, Bessey encouraged Hornaday's taxidermy as no one had done before. Billy had already expressed an interest in the art and tried to win a job at the college museum by impressing the college president with a demonstration of his skills. He shot a squirrel and prepared it in his dormitory room. With his squirrel under his arm, he ambushed the college president on the way to his office. "The stern President eyed the monstrosity suspiciously," Hornaday remembered, "chuckled in a scandalized but half-amused way, and at last gently broke the news that it was 'not good enough.'" Hornaday did not get the position at the museum.[26]

But Bessey motivated Hornaday with a good challenge. "Now, young man, we are going to see how much you know about stuffing birds," Hornaday recalled his mentor exclaiming one day. "We've got a specimen for you to try your hand on, and if you succeed in mounting it decently, you may possibly get an opportunity to work in the museum." "Show me the victim," Billy gamely replied. Bessey then revealed a large white pelican that a local farmer had donated to the college. The professor supplied a copy of John J. Audubon's seminal *Birds of North America* to serve as a model. It was the first time Hornaday had seen the great artist's stunning lifelike illustrations. Enamored with Audubon's art, he described it as "the discovery of a New World." However, it was no small challenge for Billy to model a specimen from *Birds of North America*. Over the last three years he had received scant training and had little opportunity to practice. One could not simply skin and prepare animals in one's dorm room without causing a stir. Moreover, the tools at his disposal were inadequate to the task. But "with a pocket-knife, an old misfit pair of pliers, and a smooth, flat piece of steel that had once been a file," Hornaday set to work. Taxidermists required implements more akin to the surgeons of the day, and all he had at his disposal were those of a third-rate mechanic. But despite these hardships, he did not gripe while fully understanding the stakes. The better he did, the better his chance for the future.[27]

Hornaday did better than he could have imagined, and earned the coveted position as taxidermist in the college museum with a pay rate of nine cents an hour. When not working on his coursework, the thin 5 foot 7 inch teenager with brown eyes, oval head, and already receding hair line could be seen roaming the campus or surrounding fields with a shotgun, or found at his workbench carefully preparing specimens for the museum.

The pelican had a deeper meaning to him. As Hornaday remarked twenty years later, only partly in jest, "I shaped his future, and he shaped mine at the same time." Hornaday believed that the pelican had played an important role for science and education. As a taxidermist, he had resurrected the useless and lifeless and restored it to purpose. On a more personal level, he had found once and for all that he really enjoyed the work. No longer was taxidermy slightly beyond his grasp. The pelican gave him a taste of what a career working with specimens would be like. Skinning, carving out chunks of flesh, preserving ligaments, removing organs, washing the bones with carbolic acid—none of it bothered him. Nor did the blood, fat, grease, oozy guts, or foul odors. He was in his element and knew it. Hornaday would revisit the campus several times over the course of his life and never failed to peek at the handiwork that set his life on such an adventurous course.[28]

The pelican was followed by something akin to a revelation. As Hornaday walked across the campus in 1872, he felt a sudden burst of resolve and clarity of purpose: "I will be a zoologist, I will be a museum-builder. I will fit myself to be a curator. I will learn taxidermy under the best living teachers—and I will become one of the best in that line.... This settles it! I will bring wild animals to the millions of people who cannot go to them!" Fifty-five years later, the college commemorated the event with a stone tablet marking the location of his inspiration. In a letter he wrote in 1916 to former president Theodore Roosevelt, Hornaday described how he framed the question to himself at Ames. "Shall I become a systematist and describer of species, and work for the scientific few? Or shall I devote my life to making animal and bird lore, and animal and bird specimens, *available to the millions?*" He clearly emerged from his studies with Bessey driven to succeed and convinced that the career he had chosen was of great social importance.[29]

One April day in 1873 while Hornaday worked on a contribution to the museum, Professor Bessey read a magazine article on Henry A. Ward's Natural History Establishment in Rochester, New York, to Billy as he worked. Ward's Establishment was perhaps the most influential and largest of several firms in the East specializing in collecting, preparing, and supplying specimens to the expanding museum market both in the United States and Europe. Such businesses provided many young men like Hornaday who thirsted for adventure the opportunity to learn their craft from the best in the field. Hornaday addressed a letter to Ward that night. "But my knowledge of the art is limited," he admitted after outlining his modest accomplishments in the college museum, "and it is my wish and determination to make a first-class taxidermist." Ward replied with an offer of a position as a workman, but Hornaday demurred, preferring to complete at least his sophomore year.[30]

When the term ended in November, Hornaday boarded a train bound for Rochester. "Of course I will be ready to begin immediately and stick to it," he wrote in demonstration of his determination. Any job at Ward's Natural History Establishment would be better for his career than remaining in Ames, where he had no one to tutor him in taxidermy and lacked opportunities for specimen collection and mounting. He already felt the restrictions of his self-directed study at the college museum. A college degree itself meant little to him, and Hornaday left Ames unapologetically without one.[31]

In the fall of 1873, Hornaday arrived at the doors of the world-renowned Natural History Establishment in Rochester. "THIS IS NOT A MUSEUM," a sign on the door read, "but a working Establishment where all are very busy." Nor was it a vocational school. "We were not paid to learn how to do things," Frederic Lucas wrote of Ward's; "we learned on our own time by our own efforts." Hornaday followed this established plan. "My salary is now $9.00 per week with which I am well satisfied," he wrote to Bessey soon after his arrival, "but I have to give one of the men $3.00 per week for instruction and help." Hornaday absorbed the lessons at the Establishment well and soon earned the favor of his demanding boss. He entered the Establishment a workman cleaning up gory messes left by the taxidermists and emerged nearly a decade later at the forefront of the craft. His decision to leave Ames had proven to be a wise one.[32]

Henry Augustus Ward had combined his interests in natural history and business to create his unique Establishment. As a professor of geology at the University of Rochester, he understood the booming demand for specimens in a time of unprecedented museum growth. His Establishment collected and prepared specimens as he brokered trades across continents between museums, individual collectors, and other businesses to obtain species from around the globe. Hornaday remembered Ward as a "remarkable man." "He had the nervous energy of an electric motor, the imagination and vision of a Napoleon, the collecting tentacles of an octopus, and the poise of a Chesterfield." Ward's grandson, however, presented a different portrait. He found his grandfather to be inattentive to his business, largely because his interests in academia, collecting, and natural history far superseded any desire to sit down and do the mundane work of balancing the books. The depression of 1873 hit Ward's business hard. An ominous sense of anxiety permeated the Establishment at all levels. It became a tense place to work, and every expedition had the potential to cripple the already ailing business. This would greatly affect Hornaday when he took his turn collecting specimens on the other side of the globe.[33]

Taxidermy inspired Hornaday with a purpose in life and ignited an intense, burning ambition to prove himself. After only a short apprenticeship at Ward's, Hornaday barged into his boss's office in the spring of 1874 and demanded to go to Africa. "I am going to Du Chaillu's country in west Africa on a collecting expedition for gorillas," the nineteen-year-old pluckily stated. "Is there anything in particular that I can do for you over there?" Ward was shocked by such a brazen statement (making outrageous remarks would become one of Hornaday's characteristic traits), but he did not dismiss out of hand the idea of sending his young taxidermist out on an expedition. Here was an eager and unmarried young man with plenty of chutzpah, all traits necessary for a successful collector.[34]

Hornaday was referring to Paul Du Chaillu, the French explorer and natural historian who had made his way to the interior of Africa in the 1860s and proven, to the Europeans at least, the existence of the gorilla. It was one of the last great discoveries of a large mammal species and completed a quest dating back to the Roman natural historian Pliny the Elder, who had vaguely described African apes in the first century. In the late nineteenth century, a combination of the growing professionalization of zoology, more advanced research methods, Du Chaillu's discovery, and Charles Darwin's theory of evolution led scientists all over the world to look for variations in existing species, as well as totally new animals. They were motivated as much by the thrill of scientific discovery as by the desire to earn lasting recognition for having found something new.[35]

As Ward and Hornaday debated a collecting expedition to western Africa, "far-away western relatives," as Hornaday described them, got wind of his plan and dispatched Uncle Allen Varner to restore Billy to his senses. "Now Billy," Hornaday recorded Varner telling him, "you are only nineteen, and are far too young to go to Africa afoot and alone, to do what you intend to do there. I don't want you to leave your bones in any old African jungle." Varner offered "a clear gift" of five hundred dollars and a position he could arrange at a Buffalo commercial firm with a salary of seventy-five dollars per month in exchange for Billy quitting the taxidermy business. "Well Uncle, if you feel that way about it," Hornaday recalls saying, "I cannot go on, regardless of your feelings and judgment. I will not rob you of your $500 and I am willing to make my first venture abroad in some less dangerous place, and one that you can approve." Having given up some ground, he also affirmed his career choice. Billy left it to Ward and Varner to work out a suitable alternative. After some negotiations, they settled on a shorter collecting expedition to Cuba and Florida.[36]

In October 1874, Hornaday landed in Cuba in the midst of a rebellion known as the Ten Years' War. It was not the best time for an American to be roaming around with binoculars and a gun, especially after Spanish authorities had executed several dozen American, British, and Cuban filibusters for piracy in what was known as the Virginius Affair. At one point, suspicious militiamen arrested him, but Hornaday escaped. It was a close call. "I was a bit pleased with myself because I did not become panic-stricken nor tongue-tied," he wrote, "but stood up to be shot in good form, yet meanwhile managed to reel off enough broken Spanish words to save my devoted head and my torso." Thereafter he avoided the contested areas of the island. He bagged some crocodiles, birds, turtles, and a manatee as he struggled with the heat, flies, and the language.[37]

After spending the last week of December 1874 and New Year's Day at Key West packing up specimens for shipment, he arrived in Miami on January 6, 1875. At the time of his arrival, southern Florida was a largely unpopulated region consisting of alligator-infested swamps. Within a few years of Hornaday's departure, Florida would begin its pattern of growth and development with deadly consequence to the wildlife. Each time he returned, Hornaday would mark the evolution of Miami from tiny hamlet to genuine city by noting new buildings that occupied space where he had once bagged a specimen. "I was there in 1875 when there were only three houses at Miami," he wrote of Florida in 1928, "and nothing had been spoiled. A big rattler was caught for me on the site of the Royal Palm Hotel." Many of the isolated, swampy wildernesses Hornaday hunted in 1874 were built over by the time the great Florida real estate bubble burst fifty years later.[38]

On the boat ride across the Straits of Florida, Hornaday met Chester "Chet" Jackson, a Michigan native nine years his senior. The two struck up a friendship, and Jackson accompanied Hornaday in search of specimens in the steamy swamps of south Florida. "A good fellow every way," Hornaday described his new partner to Ward, "and I think I am fortunate in falling in with him." Before departing Florida in the middle of March, the pair shared an adventure. They collected numerous species, including nearly a dozen alligators, a large crocodile, a manatee skeleton (a rare find even in those days), and several species of sharks. Exhausting work in the damp environment took its toll on his health. In early February, Hornaday suffered, as he wrote Ward, "from a fever and general breakdown, brought on, as all agree by overwork on crocodiles and alligators."[39]

Hornaday returned to Rochester following his expedition to Florida and Cuba confident that he had proven himself in his trial run as a collector. He had managed to keep his expenses low and obtained the exact specimens Ward demanded. Despite the dangers and close call in a hostile land, he had returned unscathed.

Perhaps, however, he was too cocky upon his return. Hornaday brazenly claimed to have discovered a new species of crocodile. "With that discovery," he wrote, "I nearly had a fit. A true crocodile in Florida, (or anywhere else in the United States) was then something positively unknown, and impossible." Hornaday published his claim of having discovered the new species, which he named the *Crocodilus floridanus* (Florida crocodile), in the September 1875 edition of the influential journal *American Naturalist.* "As the first true crocodile ever taken in Florida, it created a genuine sensation," Hornaday recalled. It certainly created a sensation, but not necessarily the one Hornaday implied. He was the only one who took seriously his claim of having discovered a new species; others ridiculed his naïve assertion. A writer in *Forest & Stream* magazine chided him for it, and it became apparent that he was not the first person to produce evidence that crocodiles existed in Florida either, as he had asserted. A little more than twenty years later, the celebrated zoologist and anatomist Edward Drinker Cope put the final nail in the Florida crocodile's coffin. "Mr. Hornaday supposed these individuals to represent a species different from that of more southern waters," Cope wrote in a United States National Museum report, "but I have been unable to detect any difference." Subsequent biologists categorized Hornaday's crocodile as a redescription of the *Crocodylus acutus,* a species first described by the French naturalist Baron Georges Cuvier in 1807. Some of the scientific and popular literature in the late nineteenth century and first years of the twentieth century referred to the "Florida crocodile" but still parenthetically attributed Cuvier's Latin name to the animal. Nevertheless, Hornaday steadfastly maintained his claims throughout his long life, and wrote entries for the Florida crocodile in his *American Natural History.*[40]

Hornaday worked hard preparing the crocodile, a monster measuring more than fourteen feet that he named "Ole Boss." During his expedition in the Florida swamps, he had carefully observed the crocodiles' movements and postures so that he could make the most accurate portrayal possible. This naturalistic quest for realism became a hallmark of his taxidermy. Henry Ward sold Ole Boss to the United States National Museum for $250, a sum that paid the cost of the entire expedition. It was prominently displayed at the Centennial Exposition in Philadelphia.[41]

In the fall of 1875, Ward decided to send his proven and ambitious young collector on another expedition, this time to several Caribbean Islands off the north coast of South American and to the Orinoco River Valley in Columbia and Venezuela, an area that attracted the likes of Charles Darwin and Alexander von Humboldt. Ward had received new orders for tropical species, and his dealers were out of stock. Chet Jackson accompanied "Mr. Ward's agent Mr. W. T. Hornaday," but under separate terms. Whereas Hornaday received a seven-hundred-dollar salary for the expedition, Ward agreed to pay Jackson a commission based on the specimens he collected.[42]

Before setting out for the Caribbean, Hornaday stopped in New York City, unaware that in twenty years he would make the metropolis his home. He and Jackson visited the Astor Library for a day, their sole foray into research. As an Iowan farm boy, Hornaday was rarely impressed by cities. A temperance advocate and nativist, he viewed cities as dirty, sleazy places rampant with vice, corruption, and unwelcome immigrant masses. It should not be surprising, therefore, that on his first trip to the nation's largest city, he chose to circumvent the playhouses and restaurants and journeyed instead to Brooklyn to sit in on a sermon by the famous preacher Henry Ward Beecher. Ward met with his two young employees only on the day they set sail.[43]

After a two-week voyage, during most of which Hornaday endured wrenching seasickness, the young men arrived in Barbados, where Hornaday received something of a rude awakening. It was the first time that he found himself in an environment in which he was a minority. The sheer number of black people on the island astonished him. "The darkies and niggers are everywhere," he wrote. One wonders why this would have surprised him. Basic research before the expedition should have foretold this fact. His surprise was typical, however, of how Ward conducted his expeditions. He made demands, put together vague itineraries, and dispatched his collectors almost blind into strange, new places. But wherever he went, Hornaday took notice of the people and their habits. As surprised as he was about the number of blacks on Barbados, he was equally surprised by their industriousness and entrepreneurship, a clear comment on his racial views.[44]

Hornaday spent eleven days in Barbados, where the natives "kept Mr. H very busy," as Chester Jackson recorded in his journal, before sailing to Trinidad. They hunted alligators along the Caroni River, but a combination of bad currents and some misplaced shots produced disappointing results. "We cussed the luck," a frustrated Jackson wrote. Six days later, they daringly rode a swell into a cave, shot guacharo birds in the pitch-black darkness from a violently rocking boat, retrieved their specimens, and managed to escape before the tide rose and blocked the entrance to the cave. "Glad to get out," Jackson recorded. The native guides felt similarly and refused Hornaday's entreaties to go back in. "We could not induce those men to enter that cave again," Hornaday wrote Ward with a touch of pride, "and I have since been told it is not entered with a boat more than once in two years."[45]

After two weeks in Trinidad, Hornaday and Jackson set sail for South America, getting their first glimpse of the awesome Orinoco River Delta on March 11, 1876. One of the world's great ecosystems, the delta consists of several distinct geographical zones, more than two hundred tributary rivers, and a huge basin exceeding 500,000 square miles. In spite of his expectations, Hornaday found it wanting. Ward had demanded manatee, and Hornaday inquired throughout Barbados and Trinidad on where to best hunt the prized species. Instead of a clear answer, he received a myriad of differing answers, some blatantly contradicting others. "Do not expect *too many* manatee," he wrote Ward on April 1, 1876, "but I trust you will not be disappointed in your 'Great Expectations,' and let us all hope for the best." This realization only made Hornaday try all the harder to obtain specimens. "H. is feeling poorly, worked too hard one day paddling a corial," Chet recorded in his journal. "Face badly burnt, lips peeling off, cross and irritable, but has a strong constitution and will easily pull it through," he added of his companion. "I never worked as hard in my life for anything as for specimens while in the Delta," Hornaday wrote Ward at the end of May, "and it makes me sick at heart to think of what we did *not* get." He never shed his disappointment. "I regard [South America] as the *worst* and most *unsatisfactory* on Earth for a hunter-naturalist!" he recalled decades later to Theodore Roosevelt.[46]

After tramping through the Orinoco for nearly four months, the pair split up on June 14. Hornaday remained in Demerara in French

Guyana while Jackson headed for neighboring Surinam in search of manatee.

Hornaday posted handbills throughout Demerara looking for information on manatee. He never found one, but he collected many other species, including anteaters, numerous varieties of birds, monkeys (which he ate), puma, and a sloth. Not satisfied by this haul, Hornaday wrote a lengthy and apologetic letter to Ward. "For various reasons," he wrote, "my success here has not been satisfactory to me by any means, and yet I do not see how I could have bettered it." Ironically, on the same day Hornaday penned this letter, June 24, Jackson finally secured the prized manatee. "Glory to God in the highest!" Chet exclaimed in his journal, a clear indication of how much the entire expedition rested on single species. "If H. had only been here today, how he would have enjoyed himself." They could not salvage the skin, which spoiled, but they brought back an intact skeleton. Hornaday and Jackson reunited on the centennial, July 4, 1876, and compared notes. Although Jackson found Hornaday's haul impressive, he boastfully noted in his journal, "but I had beaten him all to smash in the manatee and boa." The next day they set sail for Barbados, where they spent several days before catching a ship for home.[47]

In "Eighty Fascinating Years," Hornaday wrote that Ward "impressed on me the value of specimens, on a dollars-and-cents basis, so Chet and I were ready to do or die in order to make this expedition as worthwhile for the good professor's Rochester museum as we possibly could." He wasn't kidding about "do or die": he survived his second collecting expedition despite almost being mauled by a puma he thought was dead, and by a jaguar when he almost stumbled on it as it feasted on a tapir carcass. In another incident, he grabbed a crocodile he had shot by the tail to keep it from escaping into the water. "It wheeled with amazing quickness," he recalled, "raised himself high on his fore legs, flung his terrible jaws with their row on row of razor-edged teeth within a foot of my face, and brought them together with a blood-curdling, 'Click!'" Thankfully, Chet Jackson was on hand to issue the coup de grâce to the raging crocodile. These were exactly the kind of perils Uncle Allen had feared. Despite the dangers, Hornaday secured the very valuable specimens he sought. He experienced moments of adrenalin-packed excitement, but he got his dose of terror when he unwittingly walked into some quicksand while hunting in

Venezuela. He fired his pistol in three-shot intervals for assistance, but only kept sinking. Next he started yelling for help. As the quicksand approached his chin, he struggled to breathe as he dog-paddled to keep from going under. Then, "with all the fortuitous aid of a feature-movie production—a canoe containing two Indians appeared," he remembered. The "angels-in-disguise" rescued him and pulled him into the boat. "My nineteen year old fortitude gave way," he wrote, "and I must admit to a very youthful snivel."[48]

Ward was pleased with Hornaday's haul, even if he only secured a single manatee skeleton. In addition, he and Jackson collected alligators, caimans, turtles, piranhas and other tropical fish, an electric eel that had shocked Jackson, armadillo, bats, marmosets, ocelots, pumas, sloths, snakes, and a variety of birds, including macaws, ibises, toucans, and turkeys, among other species. Just as importantly, Hornaday kept his expenses to a minimum, going only $3.62 over the $700 budget.

Soon plans were afoot for an even larger adventure.

2

COLLECTING NATURALIST AND HUNTER

IN 1876, William Temple Hornaday embarked on a two-year adventure to Asia that few Americans of his day could have contemplated. During his earlier "two trial trips," Hornaday ("my western man," Henry Ward called him) had demonstrated his keen eye for valuable species, a willingness to take risks, and an uncanny ability to keep his expenses to the barest minimum. Unmarried, energetic, proven, and only twenty-one years old to boot, Hornaday was the right man to fill several large orders for Asian specimens.[1]

Hornaday expected to be gone for at least three years. Before departing, he took leave to visit friends and family in Indiana, Iowa, and Michigan. Among these was Josephine Chamberlain, a young schoolteacher he had met the year before in Battle Creek, Michigan. "I remember you, vividly, in your best black silk gown at the never-to-be forgotten dinner party of Emily Fellows, blessed Emily Fellows, whose hospitality gave me the opportunity to meet the finest girl on earth," he would write Josephine in 1900. "And how you loomed up above all the other girls I had ever met up to that time—or since!" Born in Connecticut, Josephine was the daughter of an army officer and homeopathic doctor who worked at the sanitarium in Battle Creek. Hornaday proposed to Josephine, who accepted. Concerned for Josephine's safety with no man in the house, he taught her to fire a handgun before departing. "It seemed to have a salutatory effect," he wrote Chet Jackson two months later from Paris, "for the burglars come no more."[2]

In October, after Hornaday had returned from his farewell tour, he and Ward settled down to business and negotiated a contract. Hornaday pledged to follow a rigorous routine. "Mr. H. agrees to give his entire time and attention to the matter of collecting with the exception of such as may be occupied at odd times in the writing of notes, letters, journals, etc., and matters of like character," the agreement stipulated. "But it is hereby understood that such matters of private interest shall in no case be allowed to interfere with the regular work of the expedition." As demanding as this stipulation might have been, Hornaday gained a concession in the form of express permission to write and publish a memoir of the expedition. But like all the other parts of the contract, this caused considerable tension. Although mutually accepted in a spirit of comity when both men sat across the table in Rochester, the terms became a sore subject when half the globe divided them. Issues such as salary, itinerary, time management, demands for specific species, and, most of all, expense funds caused frequent disagreements. While it was difficult to implement the exact terms of their arrangement given the vast distance between employee and employer, where the exchange of ideas could take months, there was an equally important personal dynamic at work. For the controlling and manipulative Ward, the future of his business and reputation rested on the ability of a twenty-two-year-old to collect high-quality specimens of animals with which he was totally unfamiliar. To the young man, he was risking his life, alone, in an unknown land.[3]

Ward joined Hornaday for the first leg of the expedition, a two-month tour across Europe and North Africa to collect specimens and gather information on Asia. Ward found numerous valuable items, but they stumbled badly when it came to collecting information. For all of their meetings at museums, universities, and the like, Hornaday and Ward had only the dimmest picture of what to expect in Asia. Once in India, Hornaday was often surprised by the dearth of collectible wildlife. Indeed, the wildlife population of India was declining in the face of British sports hunting, interior development, and population growth. Lacking accurate information, Ward frequently demanded species and in numbers that Hornaday could not possibly fulfill in the allotted time frame. Nor did they make any significant contacts in South Asia. It was Ward's good fortune that Hornaday was such a gregarious and charming young man, because it allowed him to forge connections with important people and obtain valuable leads on the locations of desired animals.[4]

While Ward met with museum officials and university professors in England, he dispatched Hornaday on side trips to obtain specimens, including to Northern Ireland's unique geological structure known as the Giant's Causeway in County Antrim. According to local legend, a giant named Finn MacCool built the causeway, consisting of more than forty thousand basalt columns. Ward, the avid geologist, had to have some of it. But what should have been a routine excursion turned into what Hornaday hyperbolically described as "an adventure ten times more dangerous than any I experienced with the head-hunters of Borneo." The local tour guides refused to let him pass, and soon others whom the young American described as "howling bog-trotters" obstructed him from chipping away their natural heritage. He persisted and obtained the specimens. Matters grew even more heated when he arrived at a cottage outside of Belfast to prepare some donkey specimens. An angry mob attacked him. "I was struck once on the back with a spade," he boastfully told Jackson, "but not hurt at all." He retreated to his cottage and barricaded the door, but the assailants kicked it in and broke all the windows. Eventually, the police restored calm and arrested four men. The enterprising Hornaday interpreted the actions of the local population not as an attempt to protect a common environmental inheritance or protest the gratuitous destruction of a valuable donkey, but instead as an uncivilized and backwards effort to impede science and progress. In his memoir *Two Years in the Jungle*, he offered the example of his trip to the causeway as evidence for the unsuitability of home rule for Ireland. Still, Finn MacCool's handiwork impressed him. "The Causeway is *wonderful* and grand, specially the former," he wrote to Josephine.[5]

Ward had already grown on Hornaday's nerves by the time they reached London, where they stayed for over a month. From Scotland south, through Liverpool, Manchester, and Sheffield, Ward nagged Hornaday about everything. "Nothing exasperates me so thoroughly as to have a man constantly watching for childish chances to find fault when I am doing my *very* best to get the work along," he complained to Chet Jackson. Ward demanded that Hornaday be at his side. "I followed him and rode with him everywhere," Hornaday vented, "waited on him for hours and he has not needed me *once* in the least." Hornaday attempted to restrain his temper, admitting to Jackson that he did not always succeed. When he could escape his employer's overbearing presence, Hornaday visited the cultural sites at the heart of the British Empire, especially the British Museum, about which he raved.

He described London as "a vast, inhospitable wilderness of brick." It was certainly a long way from Iowa farm country.[6]

After London, Ward and Hornaday crossed the channel and spent two weeks in Paris. From his earliest boyhood, Hornaday had dabbled in art and sent drawings and models to his grandparents in Indiana. As a taxidermist, he stressed sketching as a means to record wildlife activity that could then be used to accurately re-create a scene. Later in life, as director of the Bronx Zoo, Hornaday would invite artists to paint and sculpt the animals. Paris definitely appealed to his artistic nature: of his stay in the City of Lights, he wrote that it "went like a beautiful dream."[7]

After two weeks in Paris, he and Ward headed south to Italy. After crossing the Alps on Christmas Day 1876, they halted in Rome. The artwork of Italy, like that of Paris, fascinated Hornaday, and he took extra time to see such tourist attractions as the Roman Coliseum. But there was nothing comparable in the realm of natural history. "Rome is a paradise for art," he stated, "but a desert for natural history." The Italians, Hornaday remarked, did not care for wildlife unless it was "reproduced in paint or marble." He twice ventured to Mount Vesuvius and its victim city, Pompeii, to collect lava and pumice specimens. At Naples, he and Ward shipped sixteen cases of material to Rochester before boarding a steamer for Egypt.[8]

After a five-day voyage across the Mediterranean Sea, they arrived in Egypt. "Egypt is one of the grandest countries in the world for an antiquarian," Hornaday recorded, "but one of the poorest for a naturalist." Shades of Italy, in other words. Nevertheless, they procured a significant haul of geological specimens—so many, in fact, that they were kicked out of their hotel for disturbing the other guests. After "two delightful weeks in Cairo," they went to Port Said to board another steamer for a trip through the Suez Canal and down the Red Sea. Once again, the finicky Hornaday found a city he did not like, categorizing Port Said as a "dreary, dirty, and uninviting modern town." But Hornaday marveled at the Suez Canal. It affirmed his belief in the superiority of the Western races, particularly the Anglo world. On the Red Sea they made several stops, always collecting in local markets. Before stopping in Jeddah, a city Hornaday described as "both unique and beautiful," he visited the tomb of Eve, the supposed last resting place of the mother of mankind. At Jeddah, Hornaday and Ward bid each other farewell and boarded two ships headed in opposite directions: toward India for Hornaday, and toward Europe for Ward before he returned to America.[9]

Mixed emotions gripped Hornaday. On the one hand, he was more than happy to leave his imperious master behind. Throughout the trip, Ward complained about Hornaday's writing, kept him waiting, dispatched him on missions with little in the way of instructions, and then criticized his handling of the tasks. All of this foreshadowed what was to come. On the other hand, Hornaday was now venturing completely alone into an unknown land, with little knowledge of where he was going, and not one person among the tens of millions on the south Asian subcontinent waiting for him. The prospect unsettled even the self-assured Hornaday.

Hornaday was one of only four saloon-class passengers on the voyage from Jeddah to India. The others were Captain Ross, a British officer serving in the First Sikh Regiment, and his family. Certainly the presence of the Ross family alleviated Hornaday's palpable case of "the blues," with the captain's wife taking an almost maternal interest in the young American. His acquaintance with the Ross family also had a more fortuitous outcome. Like most British army officers, the captain was an experienced hunter, and he authoritatively discussed game conditions in India with Hornaday. By the time Hornaday reached India, he had a better picture of where the game could be found. More importantly, Ross promised letters of introduction to his brother, an officer in the Royal Engineers. As an engineer, he possessed excellent knowledge of India because his work in canal construction took him throughout the wild backcountry there. "My meeting with Colonel [then Captain] Ross was indeed most fortunate, as events proved, and as I look back upon it," Hornaday recorded in his memoir of the trip, "I do not see how I could possibly have accomplished what I did, without his assistance."[10]

Hornaday arrived in Bombay on February 17, 1877. Customs officials, who charged a 10 percent tariff on his expedition outfit, or gear, disgusted him. They dismissed outright his claim to exemption as a scientist. The American consul was no more interested in Hornaday's complaints than the customs service and snubbed his appeals for assistance. As if to cushion the blow to Ward, Hornaday reported that several items, including a gun, were "maneuvered through unseen" past the customs agents.[11]

A shrinking purse was not the only rude awakening Hornaday experienced in Bombay. While waiting at the customshouse, he saw a pile of elephant tusks waiting for export and determined from their length that big tuskers might be rarer than he and Ward had supposed. "I was greatly surprised at the shortness of them all," he recalled in *Two Years in the Jungle*. Yet he remained convinced that he would, in all of India, find several decent-sized elephants for American museums. He spent his first week searching for a suitable assistant and interpreter and started each morning with a visit to the city market stalls, where he purchased specimens. But he was itching to get into the field. "Am eager to get to shooting and skinning," he wrote in his first letter to Ward.[12]

Roaming the teeming city, it did not take him long to discover that he simply despised the people of India. "I am not ashamed to say that I hate the 'gentle Hindoo,'" he wrote in *Two Years in the Jungle*. As a matter of fact, Hornaday hated nearly everything about the "gentle Hindoo," from what he ate to his religion to his daily habits and mannerisms. "I believe a man could learn to eat his grandfather by a campaign in India and Ceylon," he wrote of the food. The boy from Iowa found the Indians to be lazy and dishonest. He never appreciated the work of any of the dozens of assistants who helped him during his collecting expedition. "From first to last I had no other assistance than such as could be rendered by ignorant and maladroit native servants," he wrote. Nor did he offer a kind word for the women, describing them as "almost as homely as buffaloes." Whatever tensions he created between himself and the natives, there were larger forces at play creating friction as well. The heavy hand of British imperialism, which Hornaday considered to be a progressive and humane force in a primitive land, tended not to sit well with the average Indian. Famine threatened nearly 60 million people and killed more than 4 million in the region in which Hornaday did much of his hunting. Hornaday struggled in this tense environment, and at times enflamed it with his blatantly condescending attitude toward the inhabitants.[13]

Leaving Bombay, Hornaday took a rambling 1,500–mile rail journey on an overstuffed train northeast to Etawah in the Allahabad province on the central plain, arriving on March 13. Ward wanted Hornaday to proceed to the Ganges River to collect twenty-five crocodiles in six weeks, a good species for Hornaday, as he was familiar with them from his two previous expeditions. But Ward's time frame was demanding. Captain Ross's brother, also a Captain Ross (and a major at the time Hornaday wrote *Two Years in the Jungle*), dissuaded Hornaday from wasting his time on the overhunted Ganges and directed the American to its less traveled tributary, the Jumna, instead.[14]

It was a fortuitous detour. The Jumna proved to be the place for reptiles, and Hornaday bagged twenty-six gavials in three weeks, exceeding even Ward's expectations. He admitted, though, that he was mildly disappointed that the gavials he shot tended to be small to medium in size, and none remotely approached the Florida colossus "Ole Boss." Hornaday commented that one long-distance shot was "the best shooting I have ever done with a rifle, and it was a surprise even to myself." He fondly recalled his time spent on the Jumna as "unalloyed happiness," interrupted only by the nauseating scene of half-cremated bodies floating downstream.[15]

After the gavial hunt, Hornaday accepted an invitation to join Captain Ross and his wife on a two-week hunting party about thirty-five miles south of Etawah. Traveling in grand imperial style, Ross's party included twenty-four servants, comfortable camp furniture, ample supplies of wines and spirits, delicacies of many kinds, as well as lawn tennis and other games to occupy the women while the men hunted. Hornaday took the opportunity to lecture his host on the evils of drink, but failed to convince Ross to surrender his alcohol. "But the truth is, he would never give up drinking beer and soda water as long as he stayed in India," Hornaday wrote to Josephine. "But pshaw! What's the use of trying to reform an Englishman? Hang it they *never* reform." He had only marginally better success with the game. Even though Ross generously decreed that any animal shot by anyone in the group would go to the young American, the take was slim—only a few gazelle and deer, certainly a disappointing result in the miserably hot climate where ever-present flies and constant perspiration irritated everyone. Instead of providing succor, the rare breezes felt to Hornaday "like the breath of a furnace."[16]

Hornaday stopped at Calcutta to ship his specimens to Ward before heading south to the cooler highlands. On the way he passed through Benares, a Hindu holy city noted for the thousands of protected monkeys freely roaming about. It was maddening for the hungry collector to see so many excellent specimens off-limits. "What a fine lot of monkey skins and skeletons are here running to waste!" If depriving American museums of monkey skins was not outrageous enough, Hornaday considered Hinduism a truly contemptible religion. "Their gods and goddesses are bloodthirsty and cruel monsters, guilty of adultery and incest, and some of the rites by which they are worshipped are so obscene that they can never be recorded," he wrote. "If there is a religion in existence which is destitute of even one redeeming quality, Hindooism is the one." When one compares his experiences in India to his conservation work later in life, the irony is rich. In essence, Benares was a wildlife refuge, and Hornaday—who would labor extensively to create wildlife preserves, both big and small, at the national, state, and even private levels—chafed at his inability to shoot a protected species there.[17]

Leaving the hot, dry plains of the north behind, Hornaday arrived in the cool, damp bamboo forests of the Nilgiri Hills of the south. "As to climate and natural scenery," Hornaday wrote in *Two Years in the Jungle*, "the Neilgherries [*sic*] surpass any mountain region I have ever seen, neither cold nor hot at any season, always green and fresh, and always either softly beautiful or precipitously grand." It was, however, the most exasperating leg of his entire expedition in Asia. Further from the cities and British authorities, Hornaday continued to struggle with the Indians in his employ, who stole small amounts of food and money. On several occasions, his guides displayed their lack of knowledge of the geography by getting him lost. Although he admitted that his personal servant helped him a great deal with other natives, Hornaday labeled him "a heathen" in a letter to Ward. What's more, the weather during the monsoon season was most uncooperative, and he could not stay dry during the frequent downpours. Although he later described his outfit in India as "near perfection in size and arrangement," he still traveled as light as he could and lacked enough garments to change clothes several times a day. As a consequence, wet socks and boots turned his feet raw with painful open sores.[18]

Worst of all, he suffered crippling bouts of fever. "It was an unhealthy place, and the natives warned us to get out," Hornaday reported to Ward on June 10. "Though not at all satisfied, I prepared to get out—after a stay of 13 days—and while we were packing our things the fever caught me." At one point, fever incapacitated both the hunter and his servant, leaving them prostrate and delirious on the side of the road in the pouring rain. Luckily, a British army captain happened upon them. The captain got Hornaday to a hotel and summoned a doctor. The doctor told Hornaday he was rid of the fever and would never suffer it again, but it returned within mere hours after his departure from the hotel. In fact, fever plagued Hornaday throughout the remainder of his time in India. Ever the loyal employee, Hornaday ticked off sick days and noted that this illness cost him four Sundays, his only day off. Luckily, however, the cart full of specimens that Hornaday had collected before becoming ill was recovered.[19]

At least once a week, Hornaday penned long missives laden with details to Ward. He included exact itemized expense reports to the penny spent and the number of nails purchased. Ward needed to know that his employee was working diligently on the other side of the globe and not wasting the Establishment's time and money. To Ward, Hornaday's failure might doom the Natural History Establishment's reputation. Having undertaken his own expeditions in the past into such dangerous places as central Africa, where he too battled crippling illness and less than cooperative natives, Ward was somewhat unsympathetic to Hornaday's constant complaints. Moreover, Ward demanded elephant. The entire expedition depended on it, as he frequently reminded his collector. Ward desperately sought a letter from Hornaday announcing he had killed one, preferably two or three. Instead, he got complaints about things no one could change.

On June 3, while hunting bison, Hornaday stumbled on a herd of elephants. Lacking a permit to shoot any of these choice creatures, he resisted the urge to fire on one particularly grand tusker. After exercising considerable self-restraint, he yielded to temptation when the herd crossed his path for the third time in one day. "Twice had we resisted temptation, but here it was once more," he wrote in his memoir. "I determined to kill that largest tusker then and there, if possible, and take the consequences." But his shot failed to bring the beast down. "Then I regretted my folly in firing at the elephant and wounding a noble animal to no purpose, and likewise rendering myself liable to a fine, whether I killed him or not," he wrote. Hornaday was so bothered by his botched effort to secure the specimen that, in recounting to his employer how he crossed paths with and stalked elephants, he neglected to relate that he actually shot at one and risked the fine. "Oh! What an *aggravation!*" he wrote to Ward with more meaning than the reader could have known. He was no more honest with his fiancé. "But I dared not kill him on account of a 500 Rupee fine that hung over his head like a guardian angel," he told Josephine. James Dolph probably concludes correctly that Hornaday declined to relay the failed shot to Ward out of fear that he might anger his boss, who was already harboring doubts that he had sent the right man to Asia. At the end of June, Hornaday wrote to Ward to inform him that he was applying for a permit to hunt in the private forest of a local rajah, but it would cost ten pounds, a substantial draw on his expense fund.[20]

Despite the Nilgiri's reputation for being a renowned hunting ground, both Hornaday and Ward were disappointed by the paltry outcome there. Hornaday's extended bouts of fever certainly contributed to both their frustration and the small bag. But serendipity intervened in the form of an invitation from Albert Theobald, the British forest ranger for the Annamalai Hills. Theobald "promises *two elephants*," Hornaday exclaimed to Ward. And Theobald offered assistance on securing a permit with the rajah. With no better leads, Hornaday abandoned the Nilgiri in June and headed to the Annamalai. He took an instant liking to the forest ranger. "At the end of an hour I felt that I knew him as an old friend and comrade in arms rather than an untried stranger," Hornaday recalled of their first meeting. Theobald joined the young American on his hunt, hoping to fulfill his desire for elephant skins in a forest he knew so well. A hospitable host, Theobald shared provisions and ministered to Hornaday's recurring illness when necessary. In the middle of July, Hornaday wrote Ward to tell him, "we are *sure* of the elephants." In the meantime Hornaday collected bison, gaur, axis deer, and sundry other species, but these paled in importance compared to the elephant. Hornaday wished to assure Ward that he was following all available leads to get an elephant. Ward, however, saw only broken promises and could not understand the obstacles to Hornaday's quest for the prized species. Instead of encouraging Hornaday, Ward threatened to recall him. As an aside, Ward suggested that his collector needed his partner, Chester Jackson. Ward may have meant it sincerely or as motivation, and quite possibly both. Hornaday, however, took the suggestion as an insult, and nastily replied, "send him for *Manatee* only," a species he would not encounter in India. It was a particularly unkind comment considering that Jackson had saved Hornaday's life from a crocodile in South America. A hurt Jackson learned of the slight and demanded an explanation. Hornaday vowed he would undo "my thoughtless quotation" if he could. "But as I can't undo it," he apologized, "it only remains for me to go down on my marrow-bones and cry pardon."[21]

On July 26, 1877, Hornaday wrote Ward to inform him that the rajah had granted permission to kill an elephant and provided a labor gang to assist. This good news was offset by another bout of illness that kept him from collecting for nine full days. Discouraged by the delays, Hornaday felt cheated by the terms of the contract he and Ward had negotiated in October. "The more I think of it the more I am convinced that by the terms of our agreement I am laboring against great odds on the sickness question," a dispirited Hornaday wrote Ward. Even the British army, not known as a liberal employer, had a much fairer sick policy than Ward's Establishment. Hornaday even pondered if he would be the one writing a check at the end of the expedition to pay for all the time he lost due to illness.[22]

The rajah's preserve proved less fruitful than anticipated. Hornaday spent all of August and most of September in a vain attempt to find a tusker. He had his shots. He hit one elephant that escaped to the safe haven of the government-owned forest. Then, after four days of fever, he set out for another elephant. In yet another close call, a female charged him from fewer than twenty feet and he only had time to draw his rifle and fire without aiming. The animal veered off, and it, too, was never recovered. "It is *very* mortifying to me," he wrote of his second lost elephant. Hornaday was reluctant to fire on other species lest his rifle shots frighten away nearby elephants, but he did bag some bear, deer, and black langur monkeys, among other species. But his biggest prize, by far, was a large tiger he shot on September 25. He risked his life hunting such dangerous prey at close range on foot. Most British officers hunted the striped cat from the safety of an elephant's back. Hornaday's tiger set a record for size that stood fifteen years. The large beast even impressed the government, which bestowed almost three times the normal bounty to Hornaday in October. "Isn't it jolly," Hornaday exclaimed of this unexpected windfall. On September 15, Hornaday announced to Ward that he would be leaving the preserves soon and he apologized for getting everyone's hopes up about elephants, even if his information came from trusted sources. For the first time abroad, he seriously feared that his faraway boss and mentor would recall him to Rochester.[23]

The very next day, one of his men reported a large herd of elephants in the vicinity. This time Hornaday's bullet did the trick. "At last the tide has turned and *I've got an elephant for you,* in spite of luck, Fate, and the fever," an exultant Hornaday wrote to Ward. But there was a complication: he had poached it, a fact he concealed from Ward, again fearing his boss's reaction. While he had a permit to shoot an elephant on the rajah's forest, he did not have one to shoot an elephant in the adjacent government forest. Undeterred by the threat of a fine if he should be caught, and bending to the pressure, he had decided to take the risk of illegally killing an elephant. He swore his men to silence, plying them with tobacco, Bass Ale, and a locally made liquor. The crew worked to midnight cutting up the elephant "like so many bloody vampires," Hornaday remembered, before calling it a day. After more alcohol, they resumed in the morning. Late in the afternoon, after what Hornaday estimated as sixteen hours of labor, they had completed their task. "But at the scene of the action there was about an acre of meat, pieces of skin, blood, brains, and viscera which showed unmistakably that some great animal had been wrecked," he wrote. Convinced that this remote area would not be stumbled upon anytime in the near future, he and his crew packed up their illegal spoils and marched out of the forest.[24]

After hunting some other animals and shipping his specimens to Rochester, Hornaday experienced the lowest point of the entire Asian collecting expedition. He was dead broke, alone, and exasperated with Ward's unceasing demands. "He *never* thanks anyone, or praises or compliments *me* in the least," Hornaday complained to Jackson, "and I am told others under him fare exactly the same." In the first of several incidents of serious financial duress during the trip, Ward's bank draft disappeared and could not be located in all of India. It showed up a mere week later, but the incident shocked the ever-nervous-about-money orphan. No sooner had the bank draft appeared than Hornaday received a letter from Rochester in which Ward expanded his complaints by needling his collector about spending too much time with the pen and not enough with the gun. "My rule is to work at *all* times when I can be doing anything worth doing," Hornaday tried to assure his employer. This new complaint followed on the heels of a now-familiar threat of recall. "I don't see what is the problem with this enterprise!" Hornaday hotly replied, adding, "if this trip has not paid cost, I will take my oath that *no* collecting trip ever will."[25]

His mood vastly improved in early November, however, when he received a permit to kill an elephant in the government forest. "I have not had anything make me so happy all at once since coming to India," he wrote Ward. "Prospects of elephants never so bright as at present time." He was so confident of success that he asked Ward to post notices of his receipt of permission to shoot elephants in the government forest in some prominent American scientific magazines.[26]

Within a week of writing to Ward, he downed his second elephant. Ecstatic at his good fortune, the abstemious Hornaday jumped on the carcass and cracked open a bottle of Bass Ale to celebrate. But if his first kill had been complicated by its illegality, the second was confounded by the fact that he had been permitted to kill a male, or tusker, and he had mistakenly bought down a female. He covered up this embarrassing fact in *Two Years in the Jungle*. The elephant was also pregnant, and he recovered the fetus, another fact he obviously omitted from his published account. To compound his woes, his coolies struck, leaving him alone to work on the huge carcass while another bout of fever set in. "It was a regular strike, from sheer laziness, and was perfectly exasperating," Hornaday wrote. Only by withholding their rice ration did he entice his hungry men back to work.[27]

Hornaday returned to Madras in December. While packing up his elephant skins to send to the Natural History Establishment, he suffered the worst fever of the entire Asian journey and spent nearly two weeks prostrated. He spent a lonely Christmas, ill, and shortly after New Year 1878 caught a boat to Ceylon for the next leg of the expedition.[28]

In the second week of January, Hornaday crossed the Palk Straits and landed in Colombo, Ceylon. It was a miserable voyage, and his fever had not entirely abated. Once ashore, Hornaday summoned a British army doctor, who supplied him with a formula of quinine, sulfur, cardamom, and strychnine to take three times a day. The concoction miraculously cured Hornaday of the fever that had prostrated him on sixteen separate occasions in India. Never again would fever cripple him, and for that reason he exclaimed in gratitude in his memoir, "Ceylon holds a mortgage on my affections which will never be lifted in this world." He avoided the jungle in Ceylon, per the doctor's orders. Ward's shopping list for the Spice Island included many aquatic and geological specimens, items that kept Hornaday close to the coast. "Ceylon *is* a rich field!" he happily informed Ward. He collected frogs, reptiles, lizards, turtles, tortoises, shells, and rays. When he moved inland, he bagged crocodiles and foxes. Hornaday liked Colombo and considered it the most tolerable city in the Far East.[29]

But Ceylon had its drawbacks. A letter waiting from Ward demanded that he kill six tigers in ten days, an absurd order. "Ye gods and little fishes!" Hornaday wrote to Ward. "When I read that I got up and tore my hair, jumped up and down, kicked over the chairs, and table and finally jumped in the bath tub." He added, "I am calmer now."

Then his finances suffered a crippling blow when customs officials once again levied tariffs on his gear, in this case a 400 percent duty on the alcohol he used to preserve his specimens. He and Ward were under the impression from flawed research in England that nonpotable alcohol was duty-free. As in Bombay, the authorities denied his appeals, but his loud protests drew the attention of the *Ceylon Observer*, the local newspaper, and were even cited on the floor of the House of Commons on behalf of the cause to lower duties. It was cold comfort. He had no desire to reform the British Empire's trade policies, and he was now broke. "I must tell you that I'm not in a very gushing humor just now," he wrote to Jackson. His most pressing concern was "the naked fact that I haven't a single rupee to bless myself with and my board bill is due tonight."[30]

Once again his employer seemed oblivious to the real issues. Instead of receiving the needed cash, Hornaday opened letters threatening recall and severely chastising him for the poor classification and labeling of shipments. "No draft, no deliverance," Hornaday replied. He complained that Ward was issuing "a general discharge of grape and cannon, shells and solid shot, intended to rake me fore and aft." The classification complaint greatly irritated Hornaday because his boss was fully aware that his collector lacked manuals and training in geology, and was hampered by the fact that the locals did not know the scientific or Latin names of any of his specimens. "I *ought* to be more patient and charitable in the matter," Hornaday wrote to Chet Jackson. "I *know* that if Prof. calls me home before my trip is ended it will be *only* because he is forced to from lack of money to sustain it." It was a turning point for Hornaday. His missives to Rochester grew less defensive, and he took pains to let his employer know that he was working hard and had no time for writing. And he assured Ward that he was doing all that he could to keep his expenses at the lowest possible level. Still, he received more threats of recall and complaints, and continued to suffer from inadequate funds.[31]

By the middle of April, he was simply out of money and resorted to selling some species to the local museum to raise some cash. On one occasion, he sold a ray he had purchased for a mere three shillings to another collector for seventy-five dollars, writing later, "it almost cost me a fit of apoplexy to control my feelings while the bargain was being made." Realizing that such sales defeated the purpose of his expedition, Hornaday skimped by on loans from Lee, Hedges, and Co. At the end of April, though, he reached an impasse. Ward sent a draft for one thousand dollars to Singapore, but Hornaday could not access it from Ceylon and did not have the money to purchase a ticket to get off the island. The only thing he could do was to borrow more money from Lee, Hedges, and Co. In *Two Years in the Jungle*, he wrote gratefully of the company's offer to provide a loan and said that he realized that his only option was to "accept a loan from them without the ability to offer security, and coolly sail out of their reach." But, once again, his account left out an embarrassing detail. He did have a security, his collection. Thus, instead of sending it to Rochester, he shipped it to a warehouse in London. Just to let Ward know that he had borrowed the money under duress and was not living high on the hog, he notified his boss that he had put himself on half rations and was eating only enough to stay alive.[32]

Hornaday arrived in Singapore on May 23, 1878. At least there were no duties levied on his gear this time, but the public drunkenness in the city greatly disturbed him. Yet the natives reported prospects for orangutan in Borneo—second only to elephant in importance to him and Ward—as good. Hoping good news might cheer his nervous employer, Hornaday predicted it would not be unreasonable to expect eight specimens of the highly prized species. No sooner had that letter gone out, however, than he received one from Rochester writing of dire financial circumstances and warning Hornaday he might not receive his promised salary if things continued. Hornaday added up all the specimens he had collected and calculated their market values. "Really and honestly I had thought the expedition was paying a profit, or *would* when the collections were sold," he responded to Ward. "If my salary must go," he wrote, "it *must*, and I hereby bid it an affectionate and tearful adieu." He did remind Ward that he was getting married and would like to have some money for that event, but also promised to "work on just as conscientiously as heretofore, and put in *full time* when I am well, and we will go on just as though the salary was kept up right along."[33]

Constant worry over money compounded a loneliness he described as "a state of hell." He missed Josephine dreadfully, and it made him heartsick. "My God! I believe I would rather lose my left arm than again go through the fire I have *had* to go through since leaving home," he wrote his confidant Chet Jackson. "And yet I am only half through the struggle," he lamented.[34]

Hornaday spent most of June and early July hunting along the Klang River around Malacca on the Malaysian peninsula north of Singapore. "Oh! My soul," he wrote Ward about a monkey hunt in early June. "Don't you wish you had a chance to tear through all these thorns and have them tear through you, to creep, crawl, and scramble for hours and wade through mud and water half way up to your knees? Wouldn't it be fun?" Despite his playful tone he could not conceal his disappointment at failing to bag a rhinoceros, noting, "the results were simply disgusting." He did collect a boa constrictor, a tiger, ten squirrel monkeys, another elephant (although he took only its head), and lots of bats, but this was a small haul for the time and energy he spent.[35]

Ward continued to complain about Hornaday's insistence on writing a memoir. Despite Hornaday's many assurances that he only wrote on his limited private time, Ward harbored suspicions that his faraway employee was actually scribbling in his journals on company time. "I must write a book in order to express my thanks to my friends and to abuse my aggravators—my inquisitors," Hornaday replied from Singapore in the end of July 1878. Unconvinced, Ward responded to his collector's justifications simply with an order: "Give it up!" Perhaps Ward's true fear was embarrassment if some of the uglier facts of the expedition should come to light, or if his collector should categorize his boss as an "aggravator" or "inquisitor."[36]

Despite the renewed tensions between himself and Ward over his writing, Hornaday remained optimistic about the next leg of the expedition. "This is an orangutan funeral," he wrote, "and I must be prepared to go anywhere for them and fall into the last ditch!" He also assured Ward that "I shall be prepared to live well and *work hard*."[37]

To his great delight, Hornaday received the royal treatment in Borneo. It offered quite a contrast to his experience in India, where he was just one more white face among the millions of Indians who felt oppressed by their imperial overlords. Throughout his expedition, Hornaday complained about the Indians' attitudes toward him, although he never seemed able to grasp the roots of their resentments. In Borneo, on the other hand, he was a celebrity. As the rajah personally told him, the United States was the first to recognize his father as ruler, and Hornaday was the first American to ever visit the kingdom. Instead of resentment and hostility, Hornaday found hospitality, assistance, and respect. "I told the folks when I came to Sarawak that I proposed to completely snow under every other naturalist who had ever been here," he optimistically reported to Ward on August 21, 1878. "This is a fine place for collecting every way considered," he added. "Living is cheap, help is cheap, specimens *cheap* and *plentiful*."[38]

Plentiful indeed. He shot ten orangutans in five days. Within another three weeks, he had killed a total of thirty-one. By the end of September, he added five more, to bring his total to thirty-six. By the time he was done, he had bagged forty-three orangutans, greatly exceeding his forecast of eight to twelve.

Hunting orangutans was a dangerous and strenuous business. He had taken two boats and six men with him into the mangrove swamps. As they drifted through the water, they had scanned the treetops for any sign of orangutans. If Hornaday could take a shot, he would, but more often than not, the first shot did not do the job. It was not a question of his accuracy, but of the dense foliage, a rocking boat, and a fast-moving ape. After the orangutan fell from the treetops, sometimes at distances of 100 feet, he and his men had to be careful not to get hit by the falling ape or any debris. If the animal were still alive when it splashed into the water, which was often the case, the team took after the wildly thrashing, flailing ape to issue the fatal blow with a well-directed knife. On one occasion, Hornaday killed a mother ape but rescued its child. He seemed genuinely surprised that the little one expressed no gratitude and attempted to bite him. Hornaday held onto the ape, named it Old Man, and attempted to train it. The local Dyak people came from afar to see the odd spectacle of the white man and his little orange companion.[39]

While hunting along the Sadong River in Borneo, Hornaday became acquainted with the Dyaks. Although he characterized them as "simple-minded children of the jungle" in *Two Years in the Jungle*, he took a strong interest in them and their culture, and concluded that they were extraordinarily honest and peaceful. One of the first Westerners to have seen these people, he took long notes, and his detailed account of their society and culture provided future sociologists and anthropologists with a wealth of material.[40]

Although Ward wanted Hornaday to return to Singapore in October, Hornaday was stuck in Borneo until the next draft of money arrived, which was not scheduled until December. One of the ironic by-products of Hornaday's success hunting orangutans on the Sadong was the strain it placed on his finances. He now needed more help, which meant more wages and expenses. Caring for the skins and bones of the apes required costly chemicals, as did crating and shipping them to the other side of the globe. Hornaday found himself in a similar situation as he had experienced in Ceylon. Once again, he was stranded on an island without the money to get off and with barely enough to eat.[41]

In mid-December 1878, as Hornaday waited for further instructions, he inquired of Ward if Australia would be the next stop on his itinerary, a question he had been asking for months. In the same letter, he apologized for constantly whining about the lack of money and admitted that he missed his friends and family, especially so close to Christmas. In his heart, Hornaday wanted to conclude the expedition and return home. The financial handicaps of the expedition had taken their toll on his resolve. They were embarrassing and left him feeling hopeless. After spending several months with the friendly but primitive Dyaks, Hornaday was ready to leave Borneo. Ward sent enough to get him to Singapore, where he spent six weeks staying at the United States consulate, "unable to collect a single thing." He struggled being in the company of the consul's daughter, who asked him one evening before a social event to zip up the back of her dress. "Miss Studer is really pretty, by Jove she is," he informed Jackson, "and lively as a cricket from morning till night"; "Lord what a pretty figure." He admitted he did not consider her too bright, but he was far from home, worried over Josephine's admission that she enjoyed the company of another suitor while her fiancé was on the other side of the globe, and lonely. Doubtless it was a harmless flirtation.[42]

One evening at a dinner in Singapore, he met Andrew Carnegie. An immigrant from Scotland who had built a steel empire in the United States, Carnegie was traveling around the world at the time, and he took a keen interest in the young American and his pet orangutan, Old Man. "Listening to Mr. Hornaday was a source of rare pleasure tonight," Carnegie wrote in his book *Round the World*. Despite their obvious differences in social standing and upbringing, the two men shared several personality traits. Both were pugnacious and feisty, yet charming and engaging. Hornaday later called Carnegie an "ideal American" and a "model millionaire." Carnegie and his traveling companion solved Hornaday's dilemma of what to do with Old Man by taking the unique pet into their custody.[43]

This was no mere passing acquaintance. As with Theodore Roosevelt later, Hornaday turned a chance encounter into a long and meaningful friendship and an important source of support. In the parlance of the twenty-first century, Hornaday was a master networker. He enjoyed Carnegie's financial support for his conservation causes, and the Scotsman provided valuable introductions to other like-minded philanthropists. When Carnegie engaged in his vast program of library construction in the early twentieth century, he made sure the initial inventory of books included Hornaday titles. Looking back from the end of his remarkable life, Hornaday noted, "Although I never joined the Hero Worshipper's Union . . . [I am] outrageously glad that the one and only scandalously rich man of my acquaintance was so wise, so sensible, so human and so charming."[44]

After brief layovers in China and Japan, Hornaday landed in San Francisco on April 11, 1879. He took the train east, making a triumphal tour through the Midwest. At the University of Iowa in Ames, Bessey and other professors welcomed him back and requested that he deliver several lectures on his experiences in Asia. "I tell you they received me with open arms in Des Moines, especially the *State Register*," Hornaday joyfully informed Jackson. Two letters awaiting him from Ward marred his reunion with Josephine at Battle Creek, Michigan. Instead of welcoming his "Western Man" home, Ward launched an angry tirade at Hornaday, accusing the collector of having abandoned the expedition and intentionally trying to harm the reputation and financial standing of the Establishment. You have "travelled half way round the world collecting nothing," he charged. Hornaday dashed off a blistering reply. "I do not deserve any such treatment as I have just received at your hands," he replied. Their later reunion in Rochester was a frosty one. Months later Ward still referred to Hornaday's decision to return to America due to lack of funds as the "Australia Fiasco."[45]

When he returned to Rochester, Hornaday became something of a local celebrity. He gave lectures and speeches at a wide array of local clubs and societies, ranging from workingmen's groups to middle-class women's clubs. Fame had its limits, however, and it wore on his patience. "Why I positively *avoid* meeting acquaintances on the street as much as ever I can," he confessed to Chet Jackson in June 1879. "It's the same thing over and over and over endlessly." Although such encounters may have seemed monotonous, they demonstrated public interest in his story and vindicated the time he spent writing in his journals. Even before he left Asia, his tales had fascinated some of those with whom he crossed paths. "The recital of his adventures is extremely interesting," Andrew Carnegie wrote, "and I predict that some day a book from him will have a great run."[46]

In his two years in the jungle, young William Temple Hornaday had survived hostile natives, dangerous illnesses, savage wild animals, and serious money shortages, mostly on his own, and he emerged from the experience even more self-reliant and self-confident than before. After Asia, he never questioned his own abilities again. He always knew where he stood on issues, and what he believed, and he acted forthrightly. When his total absence of self-doubt dovetailed with his Adventist sense of righteousness and immediacy, it made for a powerful and determined human force, whatever the project, a man who was focused, determined, unyielding, and steadfast in pursuit of his goals. It also led him to question the motives of others (even those with whom he worked), avoid compromise, and become downright hostile to his enemies.

Interactions with the natives confirmed his belief in the supremacy of Anglo-American culture. Although he could see positive qualities in some of the ethnic groups he encountered, he still considered their civilizations backwards and in need of the strong, guiding hands of their European imperial masters to usher them into the modern world. He did not confine this view to the brown and yellow people of Asia. In his opinion, the Catholic Irish, too, required British overlords to bring them civilization.

On a professional level, the Asian expedition produced enormous outcomes. Hornaday not only shot and skinned animals, but drew them in their natural poses and recorded their habitats, food preferences, postures, and socialization. Although many of his notes were intended for his 1886 memoir, *Two Years in the Jungle*, others were later used to construct realistic taxidermy dioramas. Hornaday felt that his book and the dioramas were the best ways he could share his knowledge of Asia with his fellow Americans who would never get a chance to make the journey. His quest for realism dramatically changed the nature of taxidermy and museum presentations. No longer would animals be displayed on single pedestals as works of art. Now they would be arranged as if acting in the wild so that the viewer would get as natural an experience as possible.

Yet, considering that Hornaday later became a pioneer conservationist, it is striking that he drew no lessons at all from Asia, particularly India, about wildlife population decline. Nor did he take any notice of the conservation policies already in place. Granted, these policies were minimal and largely an inconvenience for Hornaday, who sought wildlife for scientific purposes, not sport. Nevertheless, he continuously ascribed any native advice on wildlife locations and populations to shifty racial characteristics and never attributed the paucity of game to overhunting. This is startling given that Hornaday's entire conservation ideology later focused on overhunting as the prime cause for precipitous wildlife population declines and extinction. Consumed as he was by collecting for the good of American museums, he did not contemplate the negative effects of his own hunting. He was but one hunter killing for a noble cause.

3

STUFFED AND LIVING ANIMALS

AFTER TWO and a half years of traveling abroad, Hornaday returned to Rochester, New York, in April 1879 with two ambitious goals. First and foremost, he wished to write a gripping and informative memoir of his journey on a level with the work of the French explorer and natural historian Paul du Chaillu, though he was in no rush to complete it. The second idea, which had gelled more slowly in his mind, could be implemented more quickly. It was to introduce a revolutionary new concept in museum presentation, the group display. He was unimpressed by the museums of the great imperial powers; he found their workmanship shoddy and their designs poor. Pieces of both England and India "would NEVER do for the museums of my Country!" he exclaimed. The pieces had only gotten worse as he had traveled east. In a letter to Henry Ward, he described the specimens in Singapore as "vandalism." Hornaday felt something had to be done to raise the standards of taxidermy in order to improve the educational value of museums around the world. He set out to spark this transformation.[1]

As was his wont, Hornaday wasted no time. Just a few hours after returning from Asia, he sketched out an idea for a display of multiple animals of different species acting out a scene he had witnessed in the jungle. Ward looked it over, then ordered him to "proceed!" For several months, Hornaday painstakingly built his complicated design from the inside out. He used wood to simulate bone and papier-mâché and clay for muscle. Completed in midsummer 1879 and entitled "Fight in the Treetops," the scene re-created a dramatic (some might say sensationalistic) battle between an orangutan and a gibbon in the canopy of Borneo. Hornaday was providing his audience with a snapshot of a natural struggle that they would never be able to see for themselves. "Of the comparatively few animals which do draw blood of their own kind through ill temper and jealousy, I have never encountered any more given to internecine strife than orangutans," he wrote in 1901. "Their fighting methods, and their love of fighting, are highly suggestive of the temper and actions of the human tough." He amplified this in his diorama with some red paint to depict bite marks.[2]

"Fight" was not the first group display ever constructed by a taxidermist. Charles Willson Peale earlier had a "Missouri bear group" at his famous museum in Philadelphia. The scene centered on a bruin collected during the Lewis and Clark expedition. But the bear group did not catch on, nor, in fact, did Peale's museum, though it was fairly well known in its day for possessing a mastodon skeleton and enjoying such famous patrons as Thomas Jefferson. Peale struggled to keep his business operating, and his children sold off the remaining items after his death and closed the doors. Hornaday lived in a different era, one that took its museums and natural history much more seriously, supporting them with public and philanthropic funds and valuing them as educational facilities.[3]

"Fight in the Treetops" debuted in August 1879 at the annual meeting of the American Association for the Advancement of Science. Hornaday also delivered a natural history lecture there on the habits of the orangutans of Borneo. Impressed by the lecture and dazzled by "Fight," G. Brown Goode, the driven, chain-smoking curator of the United States National Museum in Washington, offered Hornaday the position of chief taxidermist. Owing two years on his contract with Ward at the Natural History Establishment, Hornaday declined the invitation from a man he later described as "a progressive and daring museum-builder."[4]

In the month following the scientific meeting, Hornaday returned to Battle Creek, Michigan, to marry Josephine Chamberlain. Frederic Lucas, his closest friend and fellow taxidermist at Ward's Establishment, was the best man. Their marriage was an affectionate and strong bond that lasted until Hornaday's death fifty-seven years later. Hornaday's notes to Josie, addressed as "My Choicest blessing and solace of my soul," "My life's joy," "Dearest finest and best of women," and the like, spoke of the depth of his feelings. Will and Josie were a social couple who enjoyed spending time with friends and family. They also loved to read and shared common social values, such as beliefs in homeopathic medicine, temperance, and woman's suffrage. Despite his manly hunting escapades and ubiquitous testosterone-packed language, Hornaday was a feminist by the standards of the day. His Victorian sense of morality did not affect his feelings toward his wife. The same man who found a London burlesque show "disgusting" penned bawdy notes to Josie. "Now, my Love, it is against the rules of the Park to make love in the park 'officially,'" he wrote in 1902, "so I will have to ask you to call at my house. . . . Come prepared to work overtime and make up for lost time." Josie provided a safe haven from the rigors of work, and, later, bruising conservation battles. She frequently admonished her husband not to work too hard and exhaust himself, a plea that often fell on deaf ears.[5]

Their only child, Helen Ross Hornaday, was born on October 6, 1881. "The doctor said, 'Isn't that a monster?' which scared us very much," Hornaday recounted to his older half sister Mary soon after Helen's birth, "until we found he meant a monster in size but not in appearance." Hornaday was a fond and doting father to Helen, and as many parents are prone to do, thought his daughter far superior to any other child. "*We* think she can't be matched in this neighborhood," he wrote to his half brother David. The Hornadays never considered having more children. In a candid letter to Helen when she was fifteen, Hornaday described parenthood as "a great and even awful responsibility." He added, "I would rather take a sword and go into an arena to fight hungry tigers than to assume the responsibility of bringing eight children into this world, and thus becoming responsible for their happiness and success."[6]

In no way did marriage curb Hornaday's professional ambitions. Six months after his wedding, on the evening of March 24, 1880, he and six coworkers from Ward's Establishment formed the Society of American Taxidermists (SAT) for the purpose of taking their new methods to a larger audience. "In pursuance of its one grand object," Hornaday wrote, "all members of the Society agreed that our first act was to put the grand taboo on secret methods and professional jealousy." The taxidermists who formed the SAT were in step with the growing professionalization movement of the late nineteenth century, when sundry groups ranging from medical doctors to historians banded together to establish codes of conduct and standards. In the absence of state and federal regulations, some of these societies divided the quacks from the serious practitioners.⁷

The high-sounding principles of the founders, who aimed at nothing less than attaining an "acknowledged position among the fine arts" for taxidermy, resonated throughout the SAT's constitution. In relation to natural history, they wanted American museums to "lead the world in the quality of their material and be filled with life like animals instead of being storehouses of monstrosities." Considering Hornaday's observations of the quality of museum specimens abroad, this did seem an achievable goal. But despite their serious objectives, the SAT's founders could not completely close the door on the oddities that amateur taxidermists churned out. The section of the *Annual Report* in which SAT members sold their work included a category entitled "Grotesque Groups and Animals Grotesquely Mounted," which contained such novelties as squirrels playing dominos and a pair of mounted frogs dueling with swords.⁸

In mid-December 1880, the SAT held its first meeting—the showcase for its best work reflecting the new principles—in Rochester, New York. Hornaday took the medal for best in show for his fighting apes. Even this, however, did not represent the triumph of the group model. Frederic S. Webster's group of flamingos, which greatly impressed Hornaday, hardly received any notice at all. "The failure of the flamingo group to receive any recognition caused deep disappointment to all those who had watched its production with so much interest and hopeful anticipation," Hornaday wrote more than a decade later. Instead of flocking en masse to the group concept, taxidermists continued down familiar paths. Nor did Hornaday confine himself to the group model; he contributed traditional works as well. At the third annual SAT meeting held in New York in 1882, he mounted an elephant named Mungo and again won the best-in-show award. Although it was a single animal, the display was a masterful demonstration of the other techniques Hornaday espoused. In an article describing Mungo for *Science* magazine, Frederic Lucas praised Hornaday for "putting the new principles of mounting into practice." The key to the display's realism was that the skin was tight or loose fitting where it needed to be, something previous taxidermists using older methods could not reproduce. The effect clearly resulted from Hornaday's having removed the skin himself and witnessed the elephant in the wild.⁹

Despite the success Hornaday enjoyed as an SAT founding member, officer, and contest judge, and as a winner of prizes, the SAT caused him some degree of stress. "How I did *work* in those days!" he recalled to Charles Bessey in 1907. He had a nagging need to be the best taxidermist in the United States. "In one thing I am like you," he wrote to Henry Ward in 1882. "I don't propose to be surpassed by *anybody* in my line." His insatiable ambition led him to view other taxidermists as rivals and even threats instead of as friendly competitors. He was greatly disturbed when he learned that Frederic Webster, who normally worked with birds, might attempt to mount some mammals for the 1882 annual SAT meeting. "I will be content to praise his bird work to the skies and do none of it myself," he wrote of Ward, "but if he aspires to mammals he will find me turning my attention to birds *instantly.*" This flash of jealousy and spirited rivalry departed sharply from the good-natured and friendly professional competition Hornaday preferred to depict to the public, as well as to his private correspondents. But he was not the only one who let his ego interfere with the benevolent objectives of the SAT, as Dave Madden argues in *The Authentic Animal.*[10]

The second annual SAT meeting, held in Boston in 1881, was a flop because of what Hornaday called "the almost total lack of interest in the Society manifested by the people of that autocratic city." With the disappointing results of 1881 a recent memory, the subsequent meeting assumed great importance in Hornaday's mind: a success could revive the organization, whereas another disaster might condemn it to total obscurity. It was obvious that only New York City provided the visibility the society required, but the five-hundred-dollar cost at a time when the organization suffered, as Hornaday later stated in the *Annual Report,* "temporary financial embarrassment" presented a serious obstacle. Hornaday asked Ward for the money, and, not surprisingly, the professor demurred. Hornaday then asked Ward to put up two or three hundred dollars and ask Andrew Carnegie for the remainder. He even cheekily offered to give Ward the steel tycoon's address. But Ward balked at begging for money and refused. In October, Jacob H. Studer, an ornithologist and author of several works on birds that Hornaday particularly admired, stepped forward and offered to put up the cash, but the two clashed over important details such as when and where to have the show. After two months of constant argument, Studer withdrew his support. In a pattern that Hornaday would follow throughout his life, he attacked those he felt had double-crossed him. In a letter written to Ward after Christmas 1882, Hornaday described Studer as "a mean, childish man" and a "miserable, lying scoundrel." "I would rather *die* almost than stoop to Studer," he added. Having no show would be worse than a bad show, a realization that caused Hornaday to summarize the position of the SAT at this moment in the *Annual Report* as "threatened with a disaster which would have been almost overwhelming."[11]

In the spring, with planning for the annual meeting already months behind schedule, Hornaday decided to test the waters with his friend from Singapore. In March, he asked Andrew Carnegie for the five hundred dollars needed to fund the meeting. One senses that Hornaday was at first a little embarrassed to go hat in hand, but Carnegie cheerfully granted the request with the comment, "Is five hundred dollars all you need?" "I will give you that, and you need not bother to return it!" he added, according to Hornaday's memory. Carnegie's attitude would make it easy for Hornaday to ask him for money in the future: the steel baron would contribute regularly to Hornaday's conservation efforts and provide entrance to his fellow wealthy philanthropists.[12]

With Carnegie's money, the SAT rented out Lyric Hall in New York City for the first week of May 1883 for their third meeting. Daniel Carter Beard, future cofounder of the Boy Scouts, served as a judge. It was the first time Hornaday and Beard met. Over the ensuing decades they became good friends and allies in everything from politics and conservation to training American youth in the Boy Scouts. Hornaday described the 1883 annual meeting as the "last and best" of the SAT. It was a bittersweet comment reflecting his deep satisfaction at having salvaged a desperate situation that nevertheless failed to revive the ailing organization.[13]

As Hornaday had worked on the SAT's third meeting, his contract with Ward had expired, and he gleefully accepted the position as the chief taxidermist of the United States National Museum, a branch of the Smithsonian Institution. Hornaday deeply appreciated the experiences and lessons that Ward had provided him. "For me and my fortunes," Hornaday wrote in "Eighty Fascinating Years," "it was a most fortunate thing that Professor Ward lived and wrought, that I was able to find a place directly under his rigid influence, and that for seven busy years he was satisfied to have me as an assistant." Hornaday had stepped out of Ward's shadow only to stand at the pinnacle of the taxidermy profession.[14]

On March 16, 1882, he reported to work as the chief taxidermist of the United States. He considered it a demonstration of the SAT's success in reforming the profession. "The first really startling outcome of our fierce campaign for taxidermy uplift and the betterment of museums," he wrote, "came in 1882 when I was appointed Chief Taxidermist of the United States National Museum." Yet the new assignment challenged his skills and taxed his patience. He inherited a motley collection of materials, the core of which was the poorly catalogued remainder of the massive display from the 1876 Philadelphia Centennial Exposition. Boxes of unlabeled skins were shoved in every nook and cranny of the building. The poor condition of the specimens only made the task of properly identifying them all the more difficult. Before he could even catalogue the inventory, he had to purchase the tools and build a makeshift workroom in a shed. They were not particularly comfortable quarters in the summer heat, but the large brick building that would become the National Museum's home was not quite ready for a taxidermist.[15]

His salary was good at $125 a month, and he purchased a nice home in a park-like neighborhood. "It is a good substantial modern brick house of 8 rooms," he described it to his half brother David, "almost new, all modern improvements, in a fine open park, a most excellent neighborhood." He even told his former employer Ward that he would never worry about money again. However, the experience of having been a financially insecure orphan dogged Hornaday the rest of his days, and he worried about money constantly. The Gilded Age, as historians have come to label the period between the end of Reconstruction in 1877 and the start of the Spanish-American War in 1898, was a period of bucket shops, self-help books, Horatio Alger rags-to-riches stories, and abundant get-rich-quick schemes. The overwhelming majority of these schemes only served to rip off the naïve well-to-do, with Mark Twain and Ulysses S. Grant being the two most famous cases. Not surprisingly, Hornaday was a little caught up in the spirit of the times. He read William Mathews's self-help classic, *Getting on in the World*, and invested what money he could. "Just now I am working at various little schemes to make money with which to take life more leisurely," he wrote to David in 1884. His investments, including real estate, never made him rich, though. Instead, his main supplemental income came from writing, something made possible by his eight-hour work schedule at the National Museum, a shorter day than he had put in at Ward's Natural History Establishment. He wrote articles for such magazines as *Christian Union, Cosmopolitan, Harper's Young People,* and *Youth's Companion.* The latter periodical contracted him as a regular contributor, and he earned eight hundred dollars for an article series entitled "Canoe and Rifle on the Orinoco." Despite its success at the time, Hornaday declined to reprint the series twenty years later. "The 'Canoe and Rifle' story was badly cut by the editors of the *Youth's Companion,* and is by no means in satisfactory shape to publish under my name," he wrote.[16]

As chief taxidermist, Hornaday acted as a clearinghouse for museums around the United States. Perhaps his waning interest in the SAT stemmed from the fact that he no longer needed the organization to influence the work of others in his field. Or perhaps, as Dave Madden suggests, the National Museum provided him with a larger stage, negating the need for the annual shows. He received correspondence from individual taxidermists and organizations seeking advice on anything from mounting animals to obtaining supplies. He issued instructional pamphlets with such titles as *Brief Directions for Mounting and Preserving the Skins of Mammals* and *How to Collect Mammal Skins for Purposes of Studying and for Mounting.* Hornaday's duties took him to regional exhibitions in places like New Orleans; he would spend weeks packing up a representative collection, placing the items on proper display for the duration of the show, and then packing them back up for the return to Washington. These excursions took him out of Washington often enough to stretch his restless legs and quench his thirst for travel.[17]

It was during his tenure as chief taxidermist that Hornaday put the finishing touches on *Two Years in the Jungle,* his memoir of the Asian collecting expedition. Hornaday took his time on the manuscript, for he understood that this book would ensure the fame he craved. Published in 1885, *Two Years in the Jungle* was an entertaining and highly engaging account of his adventures in lands few Americans would ever visit. The weighty tome was part hunting book, part travel adventure, part sociological treatise, and part natural history. He took pains to provide informed and graphic zoological descriptions, including sketches, of the animals he saw and killed. For those who wanted to know what Asia and its people were like, he provided detailed descriptions of life on other side of the globe, even if these were unflattering. Hornaday drew most of the illustrations himself. Nonetheless, it was a selective history, and one comes away from the book with the impression that Hornaday enjoyed his time overseas as some sort of extended holiday. He did not mention the constant angst over money or the continuous disagreements with Ward over his itinerary. The raw desperation of some of his letters, particularly from India, is totally absent from *Two Years in the Jungle*. Undoubtedly he omitted these details in deference to Ward. Hornaday would repeat this habit in later years, as he spun his writings to put the best face forward for his readers in the general public, even though these did not necessarily match the views he expressed in private.

His publisher thought highly of the finished manuscript. "It can not fail to be well received and we look for a large sale," Charles Scribner's Sons wrote on September 19, 1885. Reviewers agreed and raved about *Two Years in the Jungle*. "All things considered," the reviewer for *Science* wrote, "this is one of the most satisfactory books of its kind that we have seen for many a day." The section on Borneo, which stood almost alone as a firsthand account of life in such a remote part of the world, drew much praise. One magazine described his discussion of the Dyaks as "the most interesting part of the work."[18]

Two Years in the Jungle had just hit the bookstores when Hornaday's life was about to take a new and unforeseen turn. In March 1886, he read a newspaper article on buffalo slaughter that he described as a "severe shock, as if by a blow on the head from a well-directed mallet." His immediate reaction as chief taxidermist was to prepare for the worst and conduct an inventory of the buffalo skins at the National Museum. Disgust overcame him when he realized that the museum possessed not one single decent specimen worthy of commemorating such a distinctively American animal. Instead, he found two "sadly dilapidated" hides and a hodgepodge of skulls and bones. The secretary of the Smithsonian, Spencer Baird, understood the gravity of the situation and directed Hornaday to "go West as soon as possible" to collect some specimens from the last remaining bands of buffalo. When word of this mission circulated, an army captain barged into Hornaday's office at the National Museum one day and told the chief taxidermist, "Well, I hear that you are going to Montana to hunt buffalo, and I would like to bet you a hundred dollars that you don't find even one wild buffalo." Describing the moment as like a "final shower bath of ice cold water," Hornaday realized that the buffalo's plight might be even more dire than he had previously supposed.[19]

Whatever excitement a big-game hunt might have stirred in Hornaday, he had his reservations about what he was about to do. "To all of us the idea of killing a score or more of the last survivors of the bison millions was exceedingly unpleasant," he wrote later, "but we believed that our refraining from collecting the specimens we imperatively needed would not prolong the existence of the bison species by a single day." As he had done in India, Hornaday justified his actions in the name of science and public education. Culling several animals from a dying race was, in his mind, an altruistic act that benefited the American public by providing valuable and instructive museum specimens. After all, the market hunters had taken their toll and left nothing for the wider American population to enjoy. His fellow citizens would have to settle for a creative display in a museum. But at least that would allow for a glimpse into the wild life of the vanishing prairie.[20]

In May 1886, Hornaday set out for Miles City, Montana, for what he billed as the last scientific buffalo hunt. The U.S. Army provided an escort, scouts, supplies, and camp equipment for the government-sponsored expedition. Hornaday queried ranchers, cowboys, and others whose paths he crossed about the buffalo. Many opined that he was wasting his time and their tax dollars on such a foolish errand. Signs of dead buffalo were everywhere, but none of the living. "Thus, though gone, the traces of the buffalo are still thick over the land," Theodore Roosevelt had remarked the year before Hornaday's hunt. One tripped over dried, bleached bones, detached skulls, mountains of dung, dried flesh, spent cartridges, empty cans, and sundry other evidence of the buffalo hunter's camp. One cattleman, Henry R. Phillips of the LU-bar Ranch, suggested Hornaday try his luck in an area between the Missouri and Yellowstone Rivers, about thirty miles northwest of Miles City.[21]

Using the only credible tip he had received, Hornaday made haste for the area the rancher suggested and established a base camp. As a sign of good fortune, they found a juvenile buffalo bull and lassoed it. However, it baffled them that a lone emaciated kid would stray so far from its herd, and they found no signs of the larger group. Hornaday kept the little one and took it back to Washington, lodging it on a farm for recuperation, where it ate too much clover and died. Hornaday then mounted it and stationed it in front of his workroom at the National Museum. Otherwise, the results of the big buffalo hunt were greatly disappointing. He shot two buffalo, but they were shedding their winter coats and were unsuitable for mounting. That he failed to consider such shedding at this time of year before setting out west demonstrates both his haste and desperation to acquire some buffalo samples for the National Museum. In later accounts he repackaged this as an exploratory trip, or voyage of discovery. Hornaday decided to end his expedition early and return in the fall, when the buffalo would have better coats for mounting. Before leaving, he begged the locals not to kill any buffalo before he returned.[22]

On September 24, 1886, Hornaday returned west to Miles City with the intention of collecting twenty buffalo in two months. The animals would have their best coats during the fall, and he needed to get off the northern plains before winter set in. Although he arrived at around 2:00 a.m. and spent the remainder of the night on a barroom floor, that did not stop "the gassing of an old fur dealer with Mr. Hornaday on buffalo hunting," wrote one member of his team. The next morning, the small, tired group obtained mounts and provisions from Fort Keogh. Within thirty-six hours of their arrival, they were on the trail in search of buffalo.[23]

After three weeks, on October 13, they saw their first one. Over the next two weeks, Hornaday and his men killed twelve buffalo. "There is skinning & skeletonizing enough on hands [*sic*] to keep me busy several days," recorded William Harvey Brown, a twenty-four-year-old Iowan and student assistant at the National Museum who accompanied Hornaday in Montana. In one incident at the end of October, Hornaday decided to return to base camp, eight miles away, instead of skinning his quarry at night. While they slept, "a gang of coyotes in human form (Piegan Indians)" discovered the four carcasses they'd left unattended, skinned them, took the tongues, and smeared war paint on what was left. Hornaday fumed at the scene and saw it as a personal insult. He most likely misunderstood the meaning of the war paint, though. "What Hornaday interpreted as a personal slight aimed directly at white hunters," the historian Michael Punke writes, "seems clearly to have been a ritual of respect toward the buffalo." Unlike George Bird Grinnell and other sportsmen, Hornaday never held the Native Americans in high regard or romanticized their aboriginal ways. As one historian has noted, Hornaday suffered from the anger and anxiety many white Americans felt in the wake of Sitting Bull's massacre of George A. Custer and the Seventh Cavalry at Little Big Horn in 1876, and this was reflected in his views of Native Americans. In his mind, the Indians were vicious obstacles to progress, not noble protoconservationists. This latest incident only enforced his prejudices.[24]

On November 20, on schedule, Hornaday and his men killed their twentieth buffalo, two months from the day that Hornaday had departed Washington. Only a decade prior, given the numbers of buffalo at the time, twenty could have easily been dispatched in a single afternoon. Indeed, Harry Andrews, a buffalo hunter Hornaday met in Montana, had killed sixty-three in a single afternoon. In another example of excess, two brothers in 1872 killed more than seven thousand buffalo in only three months.[25]

As one of the men in his crew journeyed back to get wagons sufficient to haul the specimens back to Miles City, Hornaday and the rest of his men spent a week snowed-in in a makeshift cabin, playing poker, drawing pictures, and trading hunting stories. Once the blizzard subsided, they resumed the hunt, and Hornaday bagged the grandest buffalo of the entire hunt. "He was a perfect monster in size, and just as superbly handsome as he was big," Hornaday wrote. "In his majestic presence the finest of all our other buffalo bulls were quite forgotten." He took his time observing the dying animal, taking "one mental photograph of him after another," according to his *Cosmopolitan* magazine article, and sketching it. He considered it "an opportunity of a lifetime" to observe a stationary buffalo at close range for such a prolonged period, even if the animal was suffering at the time.[26]

The wagon teams arrived late in the day of December 12, just as the party had exhausted the last of their provisions. The following morning they packed up camp, crossed the Yellowstone, which was icing up, and returned to Miles City. Hornaday dashed off a quick note to Spencer Baird of the Smithsonian Institution informing him of the successful mission. The United States National Museum now had, according to Hornaday's estimation, "the finest and most complete series of buffalo skins ever collected for a museum, and also the richest collection of skeletons and skulls." "The results have been highly gratifying," an exultant G. Brown Goode wrote in the Smithsonian's *Annual Report*.[27]

On December 30, 1886, Hornaday reached Minneapolis, Minnesota. There he told reporters that the buffalo were indeed in a perilous position, and that his expedition had been "made just in the nick of time." From there he journeyed to Battle Creek, Michigan, where he rendezvoused with Josephine and other family members. They lavished food on him, and Hornaday reported to W. Harvey Brown, his assistant on the last buffalo hunt, that he "narrowly escaped death" from "good dinners, and from talking myself to death." His hunt was the primary topic of conversation, and everyone he met wanted to hear all about it. "Oh! It was pitiful to have to go all over that whole buffalo hunt in every house we visited," Hornaday exclaimed. Either Hornaday was "joshing" Brown, or his own little-known fear of public speaking got the better of him. Either way, the constant retelling of the story must have helped him craft the newspaper and magazine articles he later authored on the last scientific buffalo hunt. His innate gift for storytelling no doubt received confirmation in the faces of excited and eager listeners.[28]

Throughout 1887, Hornaday had one thing on his mind: buffalo. He was obsessed with them. He worked by day turning the animals he had killed into a powerful six-animal group display that the Smithsonian's *Annual Report* referred to as "about the biggest thing ever attempted by a taxidermist." It was so large that he had to place curtains around the exhibit as he worked so the general public would not get a glimpse of the unfinished model. But that did not stop an energetic New Yorker then serving as a civil service commissioner named Theodore Roosevelt from stepping behind the curtains. Roosevelt and Hornaday talked about hunting, the West, and taxidermy, three of several interests they shared. In time, Hornaday and Roosevelt would become friends, and they remained in close contact as they worked together on such matters as conservation and World War I preparedness.[29]

In the evenings, Hornaday worked on bringing the story of the buffalo's dwindling numbers to the public. This was a watershed period in his life and the start of his career as a conservationist. "Just as a care-free and joyous swimmer for pleasure suddenly is drawn into a whirlpool—in which he can swim but from which he cannot escape—so in 1886 was I drawn into the maelstrom of wild life protection," he wrote in "Eighty Fascinating Years." His first effort was a series of eight newspaper articles describing his adventure in Montana. Newspapers in Boston, Chicago, Detroit, Indianapolis, New York, Omaha, Philadelphia, Pittsburgh, Portland, St. Louis, St. Paul, and Washington carried the stories throughout March and April 1887. He also published them in *Cosmopolitan* magazine. He followed these popular writings with a technical monograph detailing the decline of the buffalo for the Smithsonian's *Annual Report* entitled "The Extermination of the American Bison." "Why did I do that?" he wrote of his report in 1935. "Because no one else was paying any attention to it!"[30]

More than a century and a quarter after its publication, "The Extermination of the American Bison" remains a powerful treatment of the subject and a valuable source on the cataclysmic demise of the buffalo. The best volume in the very slim catalogue of buffalo books available at the time, Joel A. Allen's *The American Bisons: Living and Extinct* (1876), stopped well before the hunters turned their attention to the great southern herd, the point at which the animal approached the precipice of extinction. Lacking a strong body of secondary sources, Hornaday relied heavily on primary sources, utilizing historical accounts and the personal observations of the first white men in new regions who wrote of untold millions of buffalo. To understand the perspectives of his contemporaries, he conducted personal and mail interviews and read voluminously through stacks of government reports. From barely literate cowboys and former buffalo hunters to Colonel Richard Dodge and General Phil Sheridan, Hornaday wanted to know their experiences with the buffalo and their account of its decline.

Despite its strength as an historical account, one that still provides the framework for our understanding of the tragedy on the Great Plains, Hornaday's "Extermination" had some significant flaws. Historians criticize him for his faulty numbers, harsh treatment of Native Americans, and failure to consider ecological causes behind the buffalo population's decline. He believed that Native Americans were as destructive and wasteful as the white market hunters, and without the latter's noble excuse of advancing civilization. "They were too lazy and shiftless to cure much buffalo meat," Hornaday wrote, "and besides it was not necessary, for the Government fed them." It is hard to disagree with the historian James Andrew Dolph, who wrote, "Hornaday's extreme bias toward the Indians is the most disappointing aspect of his essay."[31]

Hornaday clearly lamented the loss of the buffalo and never said, "it had to be," as other conservationists did. "We know now," George

Bird Grinnell wrote in 1925, "that the extermination was a necessary part of the development of the country." John C. Philips, another sportsman in the Boone and Crockett Club, wrote in a similar vein: "It was inevitable that the wild bison should vanish from our plains, and nothing is to be gained in bewailing their disappearance." On the contrary, Hornaday maintained that something better could and should have been done to accommodate the advance of civilization with the buffalo's right to existence.[32]

"Extermination" marked a seminal moment in Hornaday's life because it provided the ideological framework for his entire conservation career. His general analysis of the decline of the buffalo in the piece sounds almost identical to those he applied to other animals in the 1930s. His findings cemented in his mind the notion that extinction resulted from excessive hunting fueled by a desire for economic gain and accelerated greatly by technological improvements. Such improvements in transportation and firearms incalculably compounded "man's reckless greed." In another hallmark of his thought, he placed less blame on the market hunters for their greed than he did on average Americans for their apathy and unwillingness to preserve an irreplaceable national heirloom. By his estimate, fewer than 1,100 buffalo remained alive in North America in 1889, including all the Canadian buffalo, those in zoos or privately owned, those in Yellowstone, and a mere eighty-one wild animals The enormity and near-completeness of the buffalo slaughter proved to him that any and all species could be wiped out in just as short a time. "With such a lesson before our eyes, confirmed in every detail by living testimony, who will dare to say that there will be an elk, moose, caribou, mountain sheep, mountain goat, antelope, or black-tail deer left alive in the United States in a wild state fifty years from this date, ay, or even twenty-five?" he wrote.[33]

As Hornaday worked on the buffalo displays at the National Museum and his writings, another idea germinated in his mind. "It came like a thief in the night," he recalled, "when I was satisfied with life and not in the least looking for something new to conquer." It was the idea of a zoo. This was still a fairly new notion in the United States at the time. On June 25, 1887, Hornaday wrote a proposal for the creation of a modest living zoological collection and submitted it to his boss, George Brown Goode, the curator of the United States National Museum. Hornaday requested permission to assemble a small collection on a budget of five hundred dollars. He argued that he could do much with such a meager sum because of his many connections and an upcoming trip west with the U.S. Fisheries Commission that offered an excellent opportunity to acquire animals. "I have here your letter proposing a National Zoological garden for Washington," Goode told Hornaday. "It is a good idea! But Professor Baird is now a very sick man! We must lay this idea aside for the present. Later on I will take it up with you."[34]

After Baird's death two months later, Goode authorized Hornaday to move forward with his plan. On the last day of 1887, a small wooden structure that had been built to house his motley assortment of animals was opened to the public for viewing. "It immediately became quite popular with the public," a satisfied Hornaday noted in the Smithsonian's *Annual Report*. By the end of January, he had fifty-eight animals under his care.[35]

That January, Hornaday typed up another proposal, this one for a larger, grander zoological collection. This time he requested fifteen thousand dollars and sought to build a learning-focused zoo. At no point in either of his proposals, however, did Hornaday specifically mention anything about extinction or protecting a remnant herd against extermination. Instead, he emphasized education, continuing a congruous path from displaying taxidermy specimens to living animals.[36]

With his large and rapidly expanding collection, it was no longer feasible for Hornaday to care for the animals and continue to serve as chief taxidermist. On May 12, 1888, Goode appointed him curator of the Department of Living Animals of the United States National Museum. Creating a new department and convincing Congress to appropriate the funds to build a large national zoo along the lines Hornaday proposed were two totally different things, however. Over the years, Hornaday would demonstrate significant skills as a shrewd and effective lobbyist, but his initial foray into the arena proved shaky. His first tactic, to attach the appropriation for the zoo as a rider to the Civil Sundry Appropriation bill, failed, as the Democrats who controlled Congress and the White House refused to add a penny to the budget when they were campaigning for reelection on a platform of fiscal conservatism. Privately, some congressmen, even Democratic stalwarts like the former Confederate general Joseph Wheeler, encouragingly told Hornaday that he would fare better in the short congressional session after the presidential election.[37]

As Wheeler had predicted, Hornaday's efforts proved more successful in the short session, which began in December 1888 and ended in March 1889. He testified before committees, displayed a model he had made of the proposed zoo, and enjoyed the privilege of using the drawing room behind the podium of the Speaker of the House to lobby representatives. "I was the *only lobbyist* who was permitted to hunt in that otherwise inviolate sanctuary," Hornaday recalled, "and for me the open season was perpetual." Thomas Donaldson, a Philadelphia attorney with some experience lobbying Congress and writing laws, assisted him. Legislation to create a National Zoological Park with a $200,000 appropriation to purchase the land in the Rock Creek Park passed as part of the District of Columbia Civil Sundry Appropriation in the closing hours of the Fiftieth Congress in March 1889. President Grover Cleveland signed it in the last hours of his administration. The bill created a panel consisting of the secretary of interior, the secretary of the Smithsonian Institution, and the president of the District of Columbia Board of Commissioners to oversee the construction of the zoo.[38]

Hornaday set out to survey the ideally suited Rock Creek Park and lay out the zoo in practical terms. "A site almost incomparably better than any now used for this purpose in any other capital in the world," Samuel P. Langley, the new secretary of the Smithsonian, triumphantly called it in a letter to Congressman Samuel Dibble in January 1889. Hornaday spent the next several months working on the plans to turn the barren ground into a refuge for animals. By late fall, the stress of building a zoo from nothing had strained his nerves to the point that he required a break. In November, at the invitation of a cavalry lieutenant, Hornaday went hunting in Wyoming. It was a welcome escape from the rigors of work. After all, he blamed the market hunters who killed for gain, not the individual sportsman, for the near-extinction of the buffalo.[39]

Obtaining the land and laying out the zoo were obviously major accomplishments, but that solved only the immediate need. The real hard work would come in obtaining an annual operating budget to maintain the zoo and acquire a larger collection of animals. Hornaday lobbied Congress for $92,000, even though Dr. Goode was "aghast" and Langley "severely shocked" at the large sum. The House Appropriations Committee, chaired by the caustic Joe Cannon, an Indiana Republican, presented the first hurdle. As Hornaday remembered,

"Uncle Joe," as he was commonly called, remarked, "I suppose we will have to pass this damned bill!"—not exactly a ringing endorsement. Other congressmen and senators, too, found the sum a bit excessive and balked. "I regard this bill as such an outrage upon the tax-payers of this country," bellowed Congressman William Hatch, a Democrat from Missouri. With Langley in Europe in the spring of 1890, Hornaday felt that he bore an unfair share of the burden in battling a hostile Congress alone. "Once more I went to work, and throughout that campaign I had to play a lone hand," he wrote, "so far as outside help was concerned." Hornaday's feeling of abandonment colored his views of events as they rapidly unfolded over the next month. He believed he had triumphed in a Herculean task and deserved a rich reward for doing so. The secretary of the Smithsonian, however, failed to see it from that perspective.[40]

It appears that neither Hornaday nor Langley fully understood the implications of the special relationship between the District of Columbia and the federal government. Following the lead of Senator John Sherman, an Ohio Republican and brother of the famous Union general William T. Sherman, the Congress argued that the zoo fell outside the original bequest of James Smithson and should be governed by the rules of the District of Columbia under the Organic Act of 1878. At first Langley thought this was a mere political tactic of inconsequential importance, but then "reality struck," as the historian Helen Horowitz writes. Congress effectively slashed the zoo's operating budget in half and banned the zoo from purchasing animals. Langley would later write to members of the Smithsonian Board of Regents that this decision turned the zoo from "a scientific and national park" into "a local pleasure ground and menagerie." But one feature of the bill President Benjamin Harrison signed on May 1, 1890, removed the panel created by the 1889 bill and turned the management over to the Smithsonian Board of Regents, giving Langley direct control over the zoo. Hornaday wrote a scathing account of Langley and the ensuing events, recalling those of May 1890 as ones of "painful surprises and disillusionment." He eviscerated his former boss. "Professor Langley was a seasoned bachelor of lonesome habit, domineering temper, and the geniality of an iceberg," Hornaday wrote. "His 'NO!' was like the snap of a steel trap." In a letter to Hornaday on May 6, 1890, on which the recipient later scrawled "The Warning of the End" when he prepared his autobiography, "Eighty Fascinating Years," Secretary Langley informed Hornaday, "you will not permit any changes to be made . . . until the commissioners have signified their general approval of the same." Hornaday was free to make suggestions, draw out plans, and so forth, but not to actually do anything. In case the first letter was not clear enough, Langley followed up four days later with another. "I add as a general rule for your conduct as acting Superintendent, that no act of yours should put or seem to put the Secretary under any obligation to do anything."[41]

Naturally, Hornaday chafed under such rigid oversight. As a young orphan in Iowa and a lone collecting naturalist in Asia, he had learned to make important decisions without direct supervision, and he greatly enjoyed his independence. He certainly had confidence to spare in his own abilities. Hornaday requested the opportunity to prove himself on a six-month trial run. "No, Mr. Hornaday! I will-not!" Langley testily replied. He punctuated his emphatic refusal with an angry foot stomp on the carpet, according to Hornaday's memory of the meeting. But this retelling obscures Langley's main reason for reining in Hornaday. He was concerned over whether Hornaday had the requisite administrative experience or skills to manage such a large undertaking. "I could not myself feel it wise to adopt such a course in the case of a person who, whatever my confidence in his integrity and natural capacity," Langley wrote to Hornaday that May, "was without administrative experience." If James Dolph is correct in asserting that Langley's reasoning was code for "lack of degree credentials," it is the only time in his life that Hornaday suffered ill consequences for having left Iowa without a degree. In light of Hornaday's work later in the decade at the New York Zoological Park, Langley's stated (and possibly unstated) concern was completely misplaced, no matter how justifiable it seemed to him at the time.[42]

This turn of events gravely disappointed Hornaday, who resolved that he could no longer submit to such a regime. He discussed the matter with Josephine. When he asked her if he should endure these conditions, Josie exclaimed, *"Never!"* A more supportive spouse could not be found. With the backing of his family, Hornaday submitted his resignation to Langley the very next day. "I presume that after our conversation of yesterday this will cause you no surprise," he wrote with some gratification. "Since I can not have the only position in the park that would afford me the least satisfaction, there is no longer any reason why I should remain." The secretary accepted his resignation via return mail.[43]

Reflecting later, Hornaday came to believe that Langley was "an instrument chosen by the hand of Fortune to speed my progress toward the theater of my real life work." He could afford to be philosophical when he wrote this in the final decade of a long and fruitful life. But, in 1890, at the age of thirty-five, with a family to support, pressing pragmatic matters demanded his attention. Fortunately, he had a long-standing offer from friends in Buffalo to join them in a real estate venture. Having quenched his thirst for adventure and had his fill of government service, William Temple Hornaday was ready to put some money in his pockets.[44]

Hornaday packed his family and moved north to Buffalo, New York, to join the real estate business, the Union Land Exchange, as corporate secretary. The firm specialized in creating subdivisions in the rapidly expanding suburbs for the middle class and skilled artisans. Real estate seemed a sure bet in a city bursting at the seams. Property valuation had grown by 83 percent in a single decade, a business publication in 1890 proclaimed.[45]

He did not give up his busy writing schedule. Throughout the 1890s he penned a series of twenty articles for *St. Nicholas* magazine entitled "The Quadrupeds of North America," which became the core of his 1904 book *American Natural History*. He promoted natural history to American youth with a gritty version of science more familiar to John J. Audubon than to the emerging doctorally prepared, laboratory-bound scientists at Harvard, Stanford, and Yale. "The ideal way to observe and study mammals is to seek them in their haunts, field glass, note-book, and gun in hand," he wrote in an article entitled "How to Observe Quadrupeds" in *Christian Union* in 1891.[46]

Although Hornaday had retired from transforming dead animals into representations in museums, he took time to write a long-overdue definitive guide on taxidermy and collecting. From 1874 until he left his position as chief U.S. taxidermist in 1888, Hornaday had penned pamphlets and articles on specimen preparation, but there remained no single source. "In the English language there was not one manual of taxidermy of the slightest value to an ambitious worker," he wrote in later years. Hornaday had long planned to fill the void, and waited, as he told Henry Ward in 1882, only to acquire experience before he wrote his book. His term as chief taxidermist at the pinnacle of the profession gave him what further experience he felt he needed. Published in 1891, Hornaday's *Taxidermy and Zoological Collecting* proved to be as detailed and encyclopedic as he demanded. Dr. William J. Holland of the University of Pittsburgh contributed chapters on insects and entomological collecting. It described everything from how to pack for a collecting expedition to how to skin and mount a variety of animals. In its review of the work, *Forest & Stream* magazine praised the author's "artistic temperament and the passionate love of nature." Illustrative of how deeply the buffalo slaughter had shaken him, Hornaday urged restraint in the pursuit of specimens, an appeal praised by *Science* magazine.[47]

Hornaday entered the real estate business full of optimism. "Now is the time to buy, while values are comparatively low," Hornaday recommended in 1890, "four years from now will be the time to sell, and reap the harvest." He sought to share his good fortunes with family members, bringing both Calvin and Uncle Allen Varner into several complicated land deals. Their fortunes, however, did not improve. Even before the 1893 depression crippled the Union Land Exchange, it was already in trouble. Uncle Allen, who had entered these transactions reluctantly, noted that "their firm has given up and lost lots of property" after visiting his nephew in Buffalo in July 1892. For a fourteen-month period in 1893 and 1894, the firm failed to make a single sale. Instead, Hornaday made payments on Calvin's and Allen's notes and sent money west to assist them with their other obligations.[48]

During his downtime, Hornaday spent much of 1895 writing a novel that would be published serially in the *Buffalo Illustrated Express* newspaper. *The Man Who Became a Savage* depicted an American who escaped the pitfalls of modern urban life by taking refuge among the headhunters of Borneo. It was his only novel. One trade publication commented on the book's strange title, but added more favorably, "It is said to be, both in conception and execution, decidedly original and vigorous." Works favorably depicting primitive life were common during the fin de siècle period of financial ruin when many questioned fortunes of progress. Unlike the other works of the primitivism genre, *The Man Who Became a Savage* was written notably with the author's tongue firmly implanted in his cheek; Hornaday did not take it too seriously. In one area he did demonstrate his seriousness, however: his approval of women's suffrage and equal rights. "It is high time women take a hand in law-making," he wrote. "If they were half idiots they couldn't possibly make a worse mess of our laws than the men have done." He dedicated the work to his daughter, Helen. "Gave the first copy to Helen," Hornaday wrote in his journal on February 1, 1896. "It was a complete surprise to her; and she wept copiously over my shoulder!"[49]

4

DIRECTOR OF THE BRONX ZOO

SIX DAYS after New Year 1896, William Temple Hornaday opened a curious piece of correspondence from his friend Frederic A. Lucas, "asking if I have received an offer from New York," he wrote in his journal. "Have no idea what he refers to." The next day, the entire trajectory of his life changed. He received a letter from Henry Fairfield Osborn, chairman of the New York Zoological Society's Executive Committee, inquiring if he would be interested in interviewing for the position of director of the New York Zoological Park. "I know that you have for some time retired from scientific life and have taken up active business, but the opening which presents itself for renewed scientific work in this City seems to me such a promising one that I trust it may cause you to give it serious consideration," Osborn tantalizingly wrote. Hornaday seemed the most logical choice for the role, as he was the only one in the country with the experience of building a large zoo, as the Zoological Society envisioned, from scratch.[1]

Hornaday went about his business that day as usual and conducted his standard rounds of real estate meetings. Later, he penned his response. "Replied to Prof. Osborn's letter saying that in view of the magnitude of the enterprise I would consider any proposition the Society desires to make," he recorded in his journal. Hornaday boarded a train two weeks later, met with members of the New York Zoological Society in a jam-packed couple of days, and walked the grounds of Van Cortlandt Park, the first potential site for the future zoo. He returned to New York City the following week and was offered the position of park director, with a salary of five thousand dollars.²

Events, especially fund-raising, moved slowly over the next several months, and Osborn even warned Hornaday not to disengage from the Union Land Exchange too hastily. Then, on Friday, April 3, Hornaday received a telegram directing him to be in New York first thing Monday morning. The need to abruptly depart Buffalo saddened Hornaday, as he recorded in his journal on April 5. "Felt very low spirited at leaving the family, and our beautiful house." On Monday, April 6, he gave his first interviews as the director of the New York Zoological Park to reporters of the *New York Times* and *World*. He came to New York City alone, and his wife and daughter remained in Buffalo until the fall. "I feel like a thief every time I think of you up there all by your poor lone self," he wrote to Josie in October.³

From 1896 until his retirement thirty years later in 1926, Hornaday would work for a dynamic pair of wealthy New York blue bloods, Madison Grant and Henry Fairfield Osborn. Madison Grant had been the driving force behind the creation of the New York Zoological Society. A patrician New Yorker, Grant lived comfortably off the labor of his ancestors. Although a lawyer by education, he devoted most of his energies to causes closer to his heart. Today Grant is remembered mostly as the author of *The Passing of the Great Race*, a racist diatribe first published in 1916 that argued for the innate superiority of the Nordic races. It is the kind of book F. Scott Fitzgerald so easily mocked in *The Great Gatsby*. Grant, however, took his ideas very seriously, and he considered Anglo-American culture the best of the Nordic races and threatened by swarms of eastern and southern European immigrants. Throughout his life, Grant lobbied on behalf of all forms of immigration restrictions as well as for racially purifying eugenics. His loathing of immigrants led him to battle Tammany Hall, the New York City political machine run by the Irish since the early nineteenth century, under the banner of clean, honest government. Ironically, as his brother Deforest noted, Grant parlayed his support for the reformist, good government, non-Tammany Mayor William Strong into city support for his own pet project.⁴

Like Grant, Henry Fairfield Osborn was part of the New York social and financial elite. He circulated in the rarified company of the Morgans and Roosevelts. A paleontologist by training, Osborn worked hard and wrote voluminously throughout his life. As one friend noted, Osborn "laid out more work for himself than could have been completed in ten lifetimes." His published works included more than one thousand printed pieces, ranging from essays and reviews to original research in scientific journals. He authored twenty books, mostly on evolution and paleontology. "The marvelous versatility of his mind, and his amazing output in results made up a grand total of valuable service that never can be portrayed in words," Hornaday wrote, also describing Osborn as a "brilliant scholar." At the time Grant had pulled him into the leadership of the New York Zoological Society, Osborn was serving both as a professor of zoology at Columbia and on the staff of the American Museum of Natural History. In 1908, Osborn became president of the latter. Through his roles at the museum and the Zoological Society, Osborn created a close-knit zoological community in the nation's largest city around the common purposes of public education and scientific research. Osborn was a dedicated conservationist, but unlike most early adherents, he had never been a sportsman, although he liked to fish. His life's work with dinosaurs and evolution convinced him that no matter how strong a species appeared, its existence was tenuous and conservation was necessary for survival. He shared Grant's racial views and camouflaged them as science with "The Age of Man" display at the American Museum of Natural History. He also penned the preface for *The Passing of the Great Race*.[5]

Although Grant and Osborn were the two most important members of the New York Zoological Society leadership, its core consisted of Boone and Crockett Club men—in other words, sportsmen. With the advice of club cofounders Theodore Roosevelt and George Bird Grinnell, Grant skillfully supplemented the front office with numerous members of the business, political, and social elite for window dressing. The importance of the unofficial connection between the Boone and Crockett Club and the New York Zoological Society should not be underestimated. The sportsmen's interest in the zoo stemmed from their obvious stake in preserving game species from extinction. "The Society's work will not be by any means altogether local," *Forest & Stream* tellingly wrote on the front page of its New Year's 1898 edition. "Rather it will be for the whole continent." In essence, the society would be creating a "corporate extension of a private herd," as the historian Helen Horowitz later described the zoo. The society's sportsmen decided to share the facility with the common people of the city on the assumption that the plebian masses would absorb their values and learn to appreciate wildlife as well. Men of the club associated hunting with virility and martial values that they felt were critical to the national character, which was then threatened with dilution from peasant immigrants streaming in from eastern and southern Europe. No one captured this mood better than Theodore Roosevelt. "Hardy outdoor sports, like hunting, are in themselves of no small value to the National character and should be encouraged in every way," he stated as governor of New York in 1900. Although the sportsmen were correct in thinking that zoo animals would affect the views toward wildlife of zoo patrons, they totally underestimated the depth of such feelings and were dead wrong on the shape they took. The millions of zoogoers who packed picnic lunches, took subways with their families, and waited on lines did so because it was fun to see the exotic animals. They were not looking for indoctrination. Instead of inculcating admiration for Roosevelt's sporting life, zoos inspired a softer, gentler appreciation of wildlife that often advocated a vision of conservation—as expressed by Hornaday in the 1920s—at odds with the pro-hunting sportsmen.[6]

At the end of June 1896, a little over two months after becoming New York Zoological Park director, Hornaday and Josephine departed for a two-month tour of zoos in Belgium, England, France, Germany, and the Netherlands. "It is folly to found a zoological garden without a most carefully studied general plan," he wrote to members of the New York Zoological Society. He journeyed across the Atlantic Ocean to learn what he called, in the Progressive Era language of the day, the "hundreds of fixed scientific facts" about zoo construction and management. Hornaday returned at the end of August with three journals crammed with the information he would need to build a zoo to surpass them all. No detail escaped his notice: he recorded everything from gate receipts to park-bench designs. He also sketched numerous animal displays, marking the elements he favored and those he considered faulty.[7]

On the whole, the zoological gardens of Europe deeply impressed Hornaday, and he found much to emulate. To a reporter from the *Buffalo Enquirer*, he praised the "grandeur, magnitude, beauty, and . . . magnificence" of the continent's animal parks. In his official report to Grant and Osborn, he commented favorably on the cleanliness, organization, and wide assortment of animals, noting the vast improvement since his visits with Ward in 1876. In the late nineteenth century, European nations spent handsomely building the finest and most elaborate zoological collections in the world as a clear demonstration of their imperial power. The use of exotic animals to display power and domination dates as far back as the Egyptian pharaohs and kings, and emperors from the Chinese to the Aztecs followed suit. The connection between zoos and imperial power did not escape Hornaday's attention.[8]

In each of the European zoos, Hornaday learned an important lesson. In Hamburg, Germany, he saw the best-placed shade, and in Hannover he witnessed what he considered an ideal balance of trees and open space. Cologne had the "most natural-looking rock-work here to be found in Europe," he noted. Not surprisingly, considering Kaiser Wilhelm II's imperial ambitions, Berlin possessed the most magnificent architecture. Antwerp provided an example of an overall attractive layout. Rotterdam had some of the newest buildings, designed for the comfort of the animals, a novel idea at the time. Amsterdam gave Hornaday an excellent model of animal health. He talked with all the directors and met with the famous Carl Hagenbeck, the foremost progressive zoologist in the world at the time. Digesting all that he saw, Hornaday compiled a list of nine "absolute requirements" for the zoological park he was building. For his urban patrons, he sought an accessible, but still secluded park, with level ground, numerous walkways, and adequate shade for ample walking. He wanted both the buildings and the landscape to be visually pleasing, yet functional. For the animals, he wanted proper housing designed for specific species with adequate sanitation facilities. And for both his animal wards and human patrons, he wanted neatness and order. "A complete system of protection for the animals, and for the visitors," was one of the nine "absolute requirements" he presented to the New York Zoological Society's Executive Committee. The possibility of a dangerous tiger rampaging through the grounds worried him far less than hooligans provoking and agitating the animals with cigarette butts or sticks. In the coming years, his concerns over the proper conduct for zoogoers would become a source of friction between himself and millions who visited the park.[9]

Hornaday followed his report to Grant and Osborn with a presentation at the January 1897 annual meeting of the society. He brought with him the 135 sketches and 340 photographs he had made in Europe. At the business portion of the meeting, he was relieved to see that Andrew Carnegie had replaced Andrew H. Green, who resigned due to ill-health, on the Executive Committee. Hornaday had campaigned ardently to enlist Carnegie. "At first he entered a mild demurrer to my plea," he recorded in his "Eighty Fascinating Years," "but the court overruled it. It had to be." The replacement marked a victory over Green's attempt at the outset to make the park a source of political patronage and to merge it with Central Park.[10]

With Green and his futile efforts out of the way, the society was finally free to announce its choice of location. The options were limited, and Hornaday had his heart set on South Bronx Park from the start. "It is *admirably* adapted to our purpose," he wrote on May 1, 1896, and "could not be better." "My first sensation was of almost paralyzing astonishment," he exclaimed later. "It seemed incredible that such *virgin forest* . . . had been spared in City of New York until 1896!" South Bronx Park possessed all of the characteristics he felt necessary for the zoological park. There was plenty of shade, a freshwater supply, and a good mixture of level ground and differing terrains. A nearby subway station provided access for the entire city, and a freshly installed sewage line running down one side of the park alleviated his concerns over sanitation and cleanliness.[11]

Under the European system, only dues-paying members, who, as Hornaday said of the London Zoological Society members, "paid smartly" for the privilege, could enter the zoos. This seemed wrong to the director. He desired to build a popular institution that could alter the relationship between the American people and the swiftly dwindling animal population. On the other hand, he was a middle-class American living in the post–Haymarket Street bombing era and more than a little fearful of the unchecked mob, especially one consisting of eastern and southern European immigrants chattering in foreign tongues. In his report on his European tour to the New York Zoological Society's Executive Committee, he noted the unruliness of Parisians as they flocked en masse to free days at the Jardin d'Acclimation, leaving trash and disorder in their wake at the end of the day. He sought to come up with some way to accommodate both order and free days. As the *New York Times* reported that December, Hornaday suggested two free days a week coupled with adequate security. But the municipal authorities had another idea. They informed Grant that if the society expected a smooth transfer of land, the park would have to be open to the public five days.[12]

Once the city transferred the land of South Bronx Park to the society, Hornaday got to work reshaping the earth and laying the foundations for the zoological park. In most cases, his vision matched that of Grant and Osborn, but there was one aspect of the zoo design over which their views clashed sharply. Hornaday preferred utilitarian buildings of modest design that would serve the animals well and save money for the more important function of specimen acquisition. Grant, on the other hand, hired the architectural firm of LaFarge and Heins to design and construct elaborate, ornate, neoclassical buildings resembling those in Europe's imperial zoological gardens. Hornaday demanded that the architects not cut down a single tree without his consent in a vain effort to rein in their plans. He made what arguments he could to dissuade them from their grandiose designs, but he failed miserably, with some buildings costing more than twice the budgeted amount, much to his chagrin.[13]

Despite working at his typical frenzied pace, Hornaday was unable to finish the zoo for the proposed summer 1899 opening. The cold winter of 1898-99 slowed progress to a crawl. But in May 1899 the first animals arrived, and by the end of the year, 781 animals representing 179 species inhabited the zoo. As the wards got used to their new homes, Hornaday trained the keepers and other staff in all manner of animal care, from simple handling to health. William Beebe, the curator of birds, remembered these early days of the zoo "as ones of trial and perplexity as well as achievement." Some lessons were learned the hard way, as the inexperienced staff made plenty of rookie mistakes. In one incident, a python escaped. "And, is this the way to start a new Zoological Park?" Hornaday exclaimed upon hearing the news. Loss of a valuable specimen bothered Hornaday much less than the negative publicity a newspaper account might generate. "I assembled the men who had participated, lectured them on the folly and wickedness of newspapers and pointed out the depth of our disgrace if the news of that escape once got away from us," he recalled later in a magazine article. But he got lucky; his cover-up worked. "Not one newspaper got the slightest tip of that affair until it was five months old and so hopelessly cold that it could not be warmed over," he wrote.[14]

Among the achievements of Hornaday's life, the conversion of wild South Bronx Park into the most impressive zoo in the United States certainly ranks high. Herman Merkel, a keeper at the zoo, described it as nothing less than "a matter of wonder" that the work was completed for the opening in November 1899. Hornaday deserves the lion's share of the credit for this accomplishment, thanks to the sheer force of his personality and ability to render quick and reasonable judgments. "He was a decisive man," one of his curators, Lee Crandall, wrote. Another element of his success was the emphasis he placed on logistics. "Logistics are the key to success," Jeffery Stott wrote of the philosophy of the first generation of zoological builders, "never emotions, rarely ethics. Management differs only in scale, never in purpose." Hornaday exemplified the attitude Stott described; he designed the infrastructure of the zoo with such acumen that it lasted for decades. And he put all the lessons he had amassed in Europe to excellent use.[15]

When the New York Zoological Park opened to the public on November 8, 1899, the city provided two special express trains to transport three thousand invited guests from Grand Central Station in Manhattan directly to the Bronx. From there a convoy of carriages relayed them to the gate. Hundreds of others arrived by trolleys, or by the rage of the decade, bicycles. Henry Fairfield Osborn gave a ten-minute speech to commemorate the occasion. "What our museums are doing for art and natural science," he declared, "this park and its fair botanical companion up the Bronx will do for nature, by bringing its wonders and beauties within the reach of thousands and millions of all classes who cannot travel or explore." Then William Temple Hornaday formally opened the gates. Keepers, donned in freshly tailored dark-gray uniforms with green trim, greeted the guests and escorted them through the zoo. A handful of select guests received the star treatment, a guided tour from the director himself. Before the close of the year, ninety thousand people had already visited the park.[16]

The New York Zoological Park contained only three permanent buildings when it opened in 1899. Twenty-one additional buildings were constructed over the next sixteen years. Several, such as the Reptile, Lion, and Elephant Houses, are still in use a century later. The park also contained an assortment of dens, corrals, and aviaries for bears, large herd mammals, and birds. Hornaday spent considerable time managing and overseeing the plans, construction, and budgets of these buildings. He worked closely with New York City officials on permits and the extension of utility services.[17]

Drawing on his lessons from his European tour, these buildings were innovative for their placement of compatible species together in large areas reflecting their natural terrain. This contrasted with existing zoos in America that housed all the animals in uncomfortable buildings irrespective of natural groupings. Hornaday prized animal health and visitor comfort, and boasted in the 1904 *Annual Report* that patrons frequently remarked to the staff that his zoological park lacked the usual offensive monkey and ape odors.[18]

Hornaday emphasized the educational purpose of the zoo. Between 1899 and 1922, he wrote seventeen different editions of the *Popular Official Guide to the New York Zoological Park*, an expansive guide that provided detailed sketches of each of the animals a guest would see during a visit. He stressed English and common descriptions over Latin names and scientific jargon. Hornaday's vision of education included placing of animals on stage for performances. Monkeys and apes dressed themselves, rode bicycles, danced, and ate at a table like any American middle-class family would, among other spectacles. "Wild Animal performances are no more cruel or unjust than men and women performances of acrobatics," he declared in 1922. "Such performances, when good, have a high educational value,—but not to closed minds." Hornaday believed the animals also received an education during these demonstrations, and he knew his charges well enough to speak to their uniqueness. On the whole, however, he tended not to be very affectionate toward his charges. While he adhered to the common Victorian assumption that animals possessed a personality much as humans do, his descriptions of them tended to be unflattering, and his descriptions of the animals often seemed to suggest that he was running an insane asylum. "Oh!" he wrote to his wife in 1902, "Who wouldn't run a zoo, and be responsible for a grand collection of fools and brutes!" Labeling some animals as criminals and thieves, he distributed harsh discipline, including chaining and whipping. When one unruly pachyderm injured a keeper, he hired the former Ward Natural History Establishment taxidermist Carl Akclcy to execute it.[19]

Other than additions and expansion, Hornaday offered few new innovations after 1896, as he followed the original plan with few alterations. In 1908, Henry Fairfield Osborn had the temerity to suggest to the director that the New York Zoological Park adopt some of the techniques of the great German pioneer zoologist Carl Hagenbeck, who designed natural, barless bear dens, to provide a more natural exhibit. Referring to the plan as a costly and unproven experiment, Hornaday flatly refused. He argued that Hagenbeck's design impaired the educational purpose of the zoo because it placed the bears too far away from the patrons. Osborn dropped the matter in the face of the director's obstinacy. Hornaday, who frequently consulted both new and existing zoological parks, lobbied other directors to do likewise. It was not until the Denver Zoo built a bear exhibit on the Hagenbeck model in 1918 that this now-familiar design crossed the Atlantic.[20]

From its inception, the New York Zoological Society had intended to become a national force in the conservation of wild animals. Early in its history, however, Grant and Osborn prioritized construction of the zoo and the establishment of the core herds against extinction over lobbying for state or federal laws. Hornaday, on the other hand, wanted to push the envelope and move into the burgeoning field of wildlife protection. He feared that the buffalo was not an isolated case, but only the first species threatened with impending extinction. In 1897, he had mailed questionnaires to dozens of correspondents around the country soliciting their opinions on the state of bird life in their region of the country. He compiled this data into a report entitled *The Destruction of Our Birds and Mammals,* and included it in the *Second Annual Report* of the New York Zoological Society in March 1898.

Hornaday sought to produce something of a scientific study of wildlife populations, complete with statistics and percentages. At the time, no person or agency tracked this information, and there were few methods of animal population quantification. "I have taken up this matter solely because no one else has done so in a manner to suit me," he wrote to his mentor Charles Bessey, now at the University of Nebraska, "and I think the time is ripe for a grand crusade for the better protection of our birds and quadrupeds in the districts where they are so rapidly disappearing." While he collected articles, papers, and books on the subject, Hornaday relied greatly, as he had done in his report on the buffalo, on the direct observations of those with the most firsthand experience, including guides, local sportsmen's associations, and state officials. He asked his correspondents to "make and furnish a general estimate as to the abundance of bird life about him to-day in comparison with what it was ten or fifteen years ago." To provide some focus, he asked four specific, but highly subjective, questions. He wanted to know if bird life had declined and by how much, what "class of men" was most responsible, and what animals were becoming extinct in local areas.[21]

The results were devastating. In three-fifths of the country, Hornaday learned, "bird life in general is being annihilated." Nationally, bird life had declined by 46 percent over the last ten years. Some areas did well, including Kansas, Utah, Washington, and Wyoming. He declared California, North Carolina, and Oregon as hanging in the balance. But the rest of the country was a mess. Florida led the pack with a 77 percent decline, followed by Connecticut, Montana, and the Indian Territory with declines of 75 percent. Although Hornaday never questioned the results, others found his methodology wanting. Theodore Palmer, assistant chief of the Biological Survey, considered these numbers suggestive, not conclusive. "Such estimates are of course mainly a matter of opinion, but nevertheless are interesting," he wrote of the study. Others, such as the *Auk*, the journal of the American Ornithological Union, and the *New York Times* were more critical of his calculations. Despite its flawed methodology, Hornaday's report was the first attempt to compile a nationwide picture of the state of wildlife populations in the United States. Charged by Congress with the missions of cataloguing species, mapping bioregions, and studying the interrelationship between wildlife and human food supplies, the federal Division of Biological Survey (Bureau after 1905) never conducted an inquiry like the one documented in *The Destruction of Our Birds and Mammals*.[22]

Hornaday affixed blame for the decline of bird life on several groups, such as egg-hunting American boys, songbird-killing Italians and southern blacks, and Native Americans. He proposed bold solutions, including a radical three-year closed season on all birds, followed by stringent bag limits once hunting was permitted. He was not alone in attacking ethnic and racial minorities for killing the softer species of bird life, those often associated with feminine qualities, in juxtaposition to the manly, pseudo-military pursuit of virile game animals. In addition, he suggested the extermination of the English sparrow (a hardy invader species who crowded out native birds), the elimination of certain hunting practices, the banning of egg collecting, and an end to commercial killing, to name the most sweeping of his proposals. This marked a watershed in his evolving attitude toward hunting. If his study of the buffalo convinced him that market hunters and insatiable consumer demand could drive the most powerful and virile of species to the brink of extermination, *The Destruction of Our Birds and Mammals* persuaded him that overhunting threatened all forms of wildlife.[23]

Hornaday found the results of the study of bird life as powerful as those of his study of the buffalo a decade before. "The report is attracting a great deal of attention," he wrote a friend three months after it was published, "and in many localities is going to lead to practical results; all of which is extremely gratifying." Despite his sense of satisfaction, however, *The Destruction of Our Birds and Mammals* inspired little action. The New York Zoological Society published it, but did not act on it. There was little Hornaday could do himself. His employers wanted their director focused on implementing their impressive plan for growth, not engaged in conservation. After lots of needling, Grant and Osborn finally agreed to hire George O. Shields of the League of American Sportsmen and publisher of *Recreation* magazine as a lobbyist for conservation legislation. But they pulled the plug on him in 1906. No one adopted Hornaday's proposed three-year closed season on bird hunting. The study did confirm Hornaday's belief that market-driven hunting remained the primary cause of wildlife extinction. In this way *Destruction* followed and reinforced the logic of *The Extermination of the American Bison*. The confirmation of his belief was enough for Hornaday, and he never questioned it. In *Thirty Years War for Wild Life*, he wrote that the report "left little room for argument." "In effect, from 1898 down to the present hour that 'Report' has stood unshaken," he stated in "Eighty Fascinating Years."[24]

Many of the habits Hornaday established early in his tenure as director of the New York Zoological Park remained with him throughout his years there. He usually arrived in his office at about 8.30 a.m. after a short walk from his home on nearby Decatur Avenue. He then attended to emergent matters, met his staff of curators, and toured the grounds. Matters great and small fell under his purview, from the price and portion of a child's soda at the concession stand to the punishment of animals. Every morning at 11:30 he telephoned Madison Grant. His day was full of meetings, and he often put in twelve-hour days. In the early years, he also worked at home, where he felt free from interruptions. In between the meetings, he attempted to respond to his voluminous correspondence. Some of the letter writers praised or complained about their visit to the zoo, others posed questions on natural history. Occasionally, the director even received dead birds in the mail from someone who hoped he could identify the species in question. Increasingly, the subject of conservation took up most of his letter-reading and -writing time. He received numerous requests for information ranging from statistics and photographs to solicitations from other zoologists and game dealers. Hornaday felt compelled to weed out the crooks from the reputable dealers and warned other zoological directors when he suspected a trader of foul business. He departed his office in early evening, and generally worked on Saturdays. Not infrequently, Henry Fairfield Osborn and Madison Grant interrupted his only day off on Sunday for a guided tour of the grounds. Hornaday put up with this for twenty years, until finally in 1920 he asked his bosses if they could possibly arrange a visit for a day on which he was already at the zoo. By that time he no longer lived across the street from the park.[25]

Hornaday hired a first-rate staff of curators, mostly younger men who would forge their own distinctive careers in zoology. They included Raymond Ditmars, Lee Crandall, J. Alden Loring, and William Beebe. In 1903, Hornaday added the veterinarian W. Reid Blair to the staff. A stickler for animal health, Hornaday spent a significant portion of his time with Blair and often assisted with animal operations. He demanded that his curators inspect their collections before the park opened to the public each day, take care of sick animals, supervise the keepers, and submit daily written reports. He fully supported Osborn's plans for research stations overseas, and dispatched his curators on expeditions to Asia, South America, and other exotic locales. Hornaday was not above a little nepotism: he hired H. Ray Mitchell, his nephew through the Chamberlain family, at Josie's suggestion as the zoo's business manager in 1899, even though Mitchell might not have been the best qualified. "It was a shock to me when he told me [he] had never kept a set of books, save at school," Hornaday wrote his wife in 1899. "I think this will be a surprise to you, after what you told me on this point."[26]

As a boss, Hornaday was a stern taskmaster who kept his employees focused on the matters at hand. He issued numerous edicts enforcing a dress code, standards of conduct, and the like. In 1915, he even banned gossip because it was "beneath the dignity of grown men." During a New York City investigation of the zoo management in 1920, one dentist who was on friendly terms with many of the employees stated that, "everybody trembles when they see him." Although this comment probably overstated the case, there could be no doubt that Hornaday was a formidable boss. On the other hand, he also looked after his staff. The price the New York Zoological Park paid for remaining free of political control was that its employees were the lowest paid on the city payroll. Hornaday continuously sought to compensate for their low pay rate with a liberal pension system and other benefits, a lesson he had learned in Europe. "Pension them off comfortably in old age" was the advice he had received, if you wanted to get the "maximum of intelligent, faithful, and conscientious service." Hornaday succeeded in convincing Andrew Carnegie to provide the funds necessary to capitalize a viable pension fund.[27]

Hornaday looked to his 120 to 150 or so employees as the first line of defense for civilization in his model of Victorian law and order. "Quarreling, profanity, the use of obscene language, and loitering in any workshop, storeroom or basement," he wrote in a memo to all employees in January 1900, "is fatal to good discipline, and must not be indulged in any circumstance." "All persons employed in the Zoological Park will at all times exercise the utmost diligence in preventing the entrance to, or occupancy of the park, particularly its buildings, by disorderly persons," he commanded. He further directed his staff to protect the "children and ladies without escort, from annoyance and insult." Those employees who served this Victorian code received rewards, such as two gatekeepers in 1920 who assisted in the arrest of two suspicious boys found to possess stolen property. His zeal for law and order sometimes caused a tense relationship with the New York Police Department, if the director suspected the official guardians of the public peace of failing to pay enough attention to his corner of the city. In such instances, as in 1919, he threatened to arm his own men to contain crime if the local precinct would not detail more men.[28]

In 1903, the New York Zoological Park recorded its first year of 1 million visitors. While Hornaday was justly proud of the accomplishment, he was concerned about the level of natural history education he witnessed. He conceived of zoos as playing a pivotal role in countering the trend of urbanization that was divorcing Americans, particularly children, from nature. While there was only so much the zoo itself could do, Hornaday witnessed teachers displaying alarming ignorance as thousands of schoolchildren funneled through the park's exhibits. In one incident he liked to recall, a teacher who came to the cage for the orangutan, which she called the "or'ange oo'tann" despite the signage, turned to the class, told them it was a large South American bird, and asked the class to point it out. One of the children alertly pointed to the large orange ape gawking at them, but the teacher, despite the sign, said, "Oh, no my dear. It is a large bird from South America." Such incidents understandably troubled Hornaday. But perhaps teachers and the general public could be taught, he mused, if they had a suitable textbook—one he would write.[29]

Hornaday simply did not like any of the general natural history books he had read. On one side of the spectrum stood the elitist university scientists focused on invertebrates. This crowd infuriated him to no end. They transferred the study of nature from the rugged outdoors to the effeminate laboratory, where they broke interesting animals, such as elk, buffalo, and birds, down into their very uninteresting microscopic building blocks. A wild bear was an exciting animal to study, but a wild bear's alimentary canal was not so much fun. On the other end of the scale stood the notorious writers—later dubbed "Nature Fakers" by Theodore Roosevelt—like Reverend William Long, the most fantastical exemplar of the genre, who claimed to have witnessed a bird deliberately and consciously construct a cast out of mud and plant matter to mend its broken leg. In March 1903, no less an authority than America's most eminent nature writer, John Burroughs, took Long and his fellow Nature Fakers to task in the pages of *Atlantic Monthly*. Hornaday feared the effects that Long's books were having on the children who read them as nature study in the schools. In the preface to *The American Natural History*, the book he ended up writing, Hornaday noted that they were "too absurd for serious consideration; and when put forth for the information of the young, they are harmful." "All this would be highly amusing," he wrote of the Nature Fakers to President Roosevelt in 1907, "if it were not so pitifully serious to the children of the public schools, who are using Long's book, and who are taught to believe that what they say is true." *The American Natural History*, published by Charles Scribner's Sons in 1904, was written as much to rescue the study of nature from the elitists and fantasizers as it was to provide a solid account of nature depicting animals a person could see on an afternoon tramp through the woods.[30]

The large green folio sold for two dollars. Hornaday crammed the book with the best photographs he could obtain, though capturing several of the subjects on celluloid proved difficult. He rejected one alligator photograph, as he wrote to his editor Samuel Marvin, because it looked "too tame." Lost pictures delayed production of the entire book. Although the process of photographing the animals proved very expensive, and time-consuming when lost plates delayed production, Hornaday desired to lavishly illustrate *The American Natural History* by the standards of the day. He believed that most Americans would never see these animals, and that photographs would allow readers to visualize them better. On a more ideological note, Hornaday stood with those like George O. Shields who promoted the use of the camera over the gun, and who in their most optimistic moods hoped trophy seekers could be induced to collect photographs instead of mounted heads. Hornaday wanted people to emulate George Shiras III by shooting a deer taking a late-night drink with a camera instead of a gun. "A Midnight Reflection," Shiras's photograph, won awards at the Paris and St. Louis Fairs of 1900 and 1904, and proved a worthy example of the genre. *The American Natural History* promoted the ideal of photography by advertising good examples of it and by openly discussing the merits of pursuing game with a camera instead of a gun. Scribner's had reservations about paying more than two thousand dollars to print the illustrations with what was still a fairly new and relatively expensive technology. But Hornaday refused to print the book without the photographs, and, much to his relief, Andrew Carnegie once again came to the rescue, this time with a check to offset the printing costs.[31]

Hornaday began his book with the higher mammals. Ironically, the first species he introduced were primates, which were not even native to the Americas. He justified this on the grounds that they were still the most interesting because they were the most like humans and thus deserved first place. From there he traveled to mammals, then to reptiles and amphibians, and finally to birds and fishes. Hornaday synthesized existing texts with new research and his own ample experience with wildlife to construct his portraits. He summarized the mating habits of the animals, their ranges, food preferences, life cycles, gender differences, and other pertinent facts. He also included measurements and statistics concerning the largest specimens on record. The normal treatment was for two paragraphs per species. For endangered animals, Hornaday provided detailed histories explaining the animal's decline and whatever conservation efforts and propagation attempts, if any, were in place. Hornaday depicted a nature in which humans decided who lived and who did not. Instead of the struggle for survival, it was man's aggressive overhunting that had destroyed the balance of nature and driven species to extinction. These actions, to his mind, outweighed any natural causes of shifts in animal populations. In keeping with Hornaday's distaste for Latin names, he confined them to a footnote where an interested person could find them, but where they would not bother the layman.

The popular and scientific presses viewed the book in quite different ways. General magazines gave highly favorable accounts. "In every way this is an admirable book," wrote a reviewer for the *Bulletin of the American Geographic Society. Forest & Stream* felt Hornaday's book stood head and shoulders above the other natural history books on the market. "We believe that it will do much good and it deserves a wide public, among children as well as adults," it advised. Although Hornaday had not written his book for sportsmen, it served that audience well. According to the sportsmen's code, hunters should understand nature as part of their sport, and stop such poor practices as shooting a female of a species.[32]

But the scientific press, whom Hornaday had written off anyway, proved much more critical. The reviewers of both *Science* and *American Naturalist* criticized Hornaday's penchant for ascribing human characteristics to animals, especially after he had condemned the Nature Fakers in his introduction. They had a point, and in later years Hornaday would write even more anthropomorphic representations of animals. But in his mind, at least, it was not a crime to ascribe human traits to animals, as long as you assigned the correct ones. William Long was wrong to suggest that a bird could be a doctor and construct its own cast, but Hornaday always believed that animals could be criminals. The reviewers for *Science* and *American Naturalist* also found Hornaday's treatment of the animal kingdom entirely too subjective. "The treatment of the various forms varies considerably," *American Naturalist*'s reviewer wrote, "apparently to a large extent with the author's interest." *Science* was even more critical about how he combined species into orders and groups that should not have been bound together. This approach contradicted Hornaday's praise in the book's preface of "system" as the "only master-key by which the doors of Animate Nature can be unlocked." But even if critical of his method, the scientific journals saw some positive features of the book, especially since it appealed to the general reading public. They applauded, for instance, the numerous myths Hornaday tore apart. He demonstrated, for example, that bats do not get entangled in human hair, and that some bird species generally thought to be pests to farmers are actually very beneficial consumers of crop-eating insects. *American Naturalist* described the book as "an excellent work for home reading and reference" that provided "a partial substitute for those who have no opportunity to visit a good zoological park."[33]

Both general and scientific publications found the photographs to be the best feature of *The American Natural History*. "The book is profusely illustrated with drawings and photographs, most of which are exceptionally good," commented one review. Aside from published reviews, Hornaday received several letters of praise directly from schoolteachers. One wrote to tell him that "nothing could be more practical for city teachers" than his book. Hornaday's friend and benefactor Andrew Carnegie purchased five hundred copies of *The American Natural History* and donated them to libraries he had recently established.[34]

The enduring popularity of *The American Natural History* can be measured in Hornaday's sales reports from Charles Scribner's Sons. It was his perennially best-selling work, only superseded temporarily by the sale of a new book. Within six months of any new publication, however, *The American Natural History* was back on the top of his list. Charles Scribner's capitalized on this through revisions and reprints.[35]

In 1905, the New York Zoological Society entered the conservation field in a manner much more suited to Hornaday's tastes than it had so far. Using his personal friendship with President Theodore Roosevelt, Madison Grant proposed a partnership between the federal government and the society to reverse the wave of destruction of the buffalo and initiate the process of returning the buffalo to the rolling grassy hills of the Great Prairies. Not surprisingly, given Grant's interest in so-called scientific racial theories and eugenics, he considered crossbreeding "an abomination," and saw providing buffalo with expansive land as the only genuine way to preserve the animal's vitality and purity. Hornaday and the society stressed the unique partnership that the deal involved, one in which the government provided the land, but a private organization supplied the buffalo. "We feel that it is the duty of individuals to do something toward the establishment of the series of buffalo herds that should be established very soon," a public-minded Hornaday wrote to Congressman John F. Lacey of

Iowa. In describing the resultant creation of the Wichita Forest and Game Preserve, designed primarily for the benefit of the buffalo, the society boasted of its "corporate sacrifice" and "partnership agreement" with the federal government.[36]

President Roosevelt fully sympathized with the plan as no man in the White House before him would have. A renowned big-game hunter and cofounder of the Boone and Crockett Club, Roosevelt typified sportsmen's conservation values more than any other man, and boldly used his office on behalf of wildlife protection. He created the first bird sanctuary on March 14, 1903, through executive order, and added fifty more before leaving office in 1909. "The creation of these reservations at once placed the United States in the front rank in the world work of bird protection," Roosevelt unabashedly proclaimed in his *Autobiography*.[37]

Despite the conservation successes he so impressively ticked off in his *Autobiography*, however, Roosevelt did not obtain all from Congress that he requested. In his Fourth Annual Message in December 1904, for example, he asked Congress for the power to establish game preserves in the forest reserves for the purpose of protecting the large mammals "once so abundant in our woods and mountains and on our great plains, and now tending toward extinction." Congress showed no interest in this measure at all. But the coldness displayed by the representatives and senators to this proposal failed to deter the president from offering a more specified request, one meeting Grant's proposal for the establishment of the first-ever buffalo preserve in the United States. In January 1905, he requested from Congress a bill that would carve an 8,000–acre preserve out of the 60,000–acre Wichita National Forest in Oklahoma, which itself had been recently created from an Indian reservation, to restore a near-extinct species to its former habitat. The area they had in mind already had a history in early wildlife protection. In 1901, President William McKinley had created the Wichita Forest Reserve to protect a herd of Texas longhorns.[38]

Congressman John F. Lacey, one of the lesser-known figures in the history of early conservation, had sponsored bills to protect the buffalo in Yellowstone in 1894, ban the interstate trafficking of illegally killed wild game in 1900, create Wind Cave National Park in 1903, establish the U.S. Forest Service in 1905, and protect national heritage sites with the Antiquities Act of 1906. He shepherded the Wichita bill through the House. While some people, including Hornaday, praised the Iowa Republican and Civil War veteran, Theodore Roosevelt completely ignored Lacey's contributions when writing his *Autobiography*. Nevertheless, Lacey's bill sailed through the House. The Senate moved a little more slowly, and it took a personal appearance from Lacey before the Senate Agriculture Committee in March 1906 to convince the upper chamber to add the fifteen thousand dollars required to create Wichita as a rider to the annual appropriations. "I congratulate you most heartily," an exultant president wrote to Lacey. The fifteen-thousand-dollar appropriation went mostly to building a fence around the preserve as Hornaday recommended. The purpose of the barrier was to prevent the buffalo from roaming off the land, while at the same time keeping predators from entering.[39]

As the House and Senate worked on the legislation in 1905 and 1906, the New York Zoological Society dispatched J. Alden Loring, the curator of mammals at the zoo, to survey Wichita and report back on the condition of the land and its topography. Hornaday had never been there, and Loring's information would be critical for designing the layout of the preserve, a responsibility that fell to the director. Loring's study of the ground and interviews with local residents convinced him that this land had indeed once been part of the range of the great southern buffalo herd. Loring argued that a fence around the perimeter was not only feasible, but a necessary component to protect the endangered buffalo. "There is no more reason for protecting the wolves on the Reserve than there would be in allowing a band of outlaws to live there in peace while they were plundering and murdering the neighboring settlers," he reasoned.[40]

By the summer of 1906, progress on the preserve was beginning to slow. Hornaday suspected Secretary of Agriculture James Wilson of falling under the influence of Chief Forester Gifford Pinchot, who did not want to use forest for wildlife preservation. The true culprit for the delay, however, was the federal vendor contract process. Everyone agreed that the preserve should be fenced to keep out predators, and Hornaday recommended the Page Fence Company, which he believed was the only one capable of meeting the specifications. Federal requisition policy, however, required a bidding process for all contracts. Hornaday considered this an annoying delay, especially when he wrote specifications that he knew only Page could fulfill. Therefore, he was surprised to learn in mid-August 1906 that Page was not one of the seven companies that had entered a bid for the contract. Hornaday wrote that very day to Page, pressing them to submit a proposal so that the fence could be completed by the fall. This, however, was impossible, as the president of the company informed him. Even if they received the bid that day, the manufacture, transportation, and installation could not be completed before the onset of winter. The buffalo would have to wait one more year for their restoration to the plains.[41]

In October of the following year, the preserve was finally ready to receive the shipment of fifteen buffalo. Several buildings had been constructed, the fence built, and every undesirable species inside the fence hunted down. Hornaday designed a special crate to house each buffalo in relative safety and comfort during its long rail journey from the Bronx to Oklahoma. In keeping with the theme of cooperation and the bestowal of a gift, the New York Central Railroad and the American Express Company transported the animals from New York to St. Louis free of charge. From the Gateway City, the Wells Fargo Express Company conveyed them to Cache, Oklahoma, for free as well.[42]

Congressman Lacey fretted over the threat of Texas fever, a fatal tick-borne illness that had decimated the cattle population of Oklahoma, but Hornaday brushed those criticisms aside. "I feel sure that we are doing right in making a trial of the Wichita Forest Reserve," he wrote Lacey, "because it is so absolutely ideal in every respect otherwise than the naked fact that it is within the boundary of the Texas fever area." A careful investigation found no evidence of ticks inside the reserve. When the time came a year later to release the herd on the plains, the cowboys handling the buffalo coated them in petroleum to prevent ticks from entering the reserve. This prophylactic proved decisive and saved all but one animal. By 1913, there would be thirty-eight buffalo living on their ancestral ground.[43]

The opening of the Wichita Forest and Game Preserve was perceived to be an important event at the time, and Hornaday received praise for his efforts. "Director Hornaday of the Bronx Zoological Park deserves the gratitude and encouragement of the Nation as the chief preserver from extinction of the American Bison," editorialized the *New York Times*. The preserve was equally important historically. "It was the first managed wildlife preserve or refuge in the United States," the historian Gerald Williams points out. But it only whetted Hornaday's appetite for more.[44]

William Hornaday Senior, 1850s.

WILLIAM T. HORNADAY COLLECTION, PRINTS AND PHOTO-
GRAPHS DIVISION, LIBRARY OF CONGRESS, LC-DIG-DS-02104

William Temple Hornaday (center) at about age six
with his half brother Calvin and half sister Mary.

WILLIAM T. HORNADAY COLLECTION, PRINTS AND PHOTO-
GRAPHS DIVISION, LIBRARY OF CONGRESS, LC-DIG-DS-02103

William Temple Hornaday (center) and his assistant at the model and taxidermy shop of the United States National Museum in the 1880s.

COURTESY OF THE SMITHSONIAN INSTITUTION

Opposite page
Examples of Hornaday's revolutionary taxidermy techniques: skeleton of an American bison (*top*), and mannikin for male American bison, half finished (*middle*) and completed (*bottom*).

FROM HORNADAY AND HOLLAND'S *TAXIDERMY AND ZOOLOGICAL COLLECTING*, 10TH EDITION [1912], FACING PP. 298, 152, AND 156

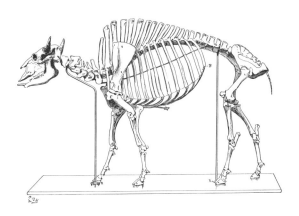

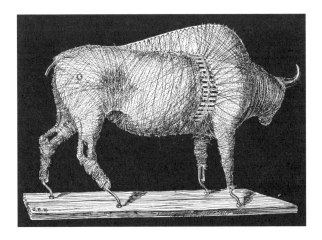

The fruits of the last buffalo hunt; the famous group of American bison, collected and mounted by Hornaday, in the National Museum.

Opposite page
Smithsonian Secretary Samuel P. Langley and William Temple Hornaday (*top*, pictured together at center) surveying the grounds of the newly established National Zoological Park, circa 1888, with a group that included Frederick Law Olmsted. Schoolchildren (*bottom*) viewing the first bison at the National Zoological Park in 1889.

COURTESY OF THE SMITHSONIAN INSTITUTION

William Temple Hornaday, Josephine Chamberlain Hornaday, and Helen Ross Hornaday in the 1890s.
COURTESY OF STEPHEN HAYNES, MINNEAPOLIS, MINNESOTA

A herd of buffalo in Bronx Park, circa 1902, from the City of New York Annual Report for the Year of 1902.

COURTESY OF NEW YORK CITY PARKS

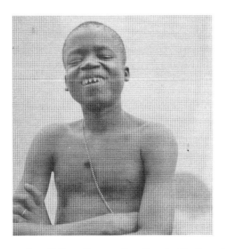

Photograph of Ota Benga by Jessie Tarbox Beals, 1904.

COURTESY OF THE ST. LOUIS PUBLIC LIBRARY

Hornaday at his desk at the Bronx Zoo, circa 1910. Note the mounted heads and the names of famous naturalists on the walls.

COURTESY OF THE WILDLIFE CONSERVATION SOCIETY

Opposite page
William and Josephine at their home, the Anchorage, in Stamford, Connecticut, in 1924 (*top*); and the face of an old curmudgeon—Hornaday around 1925 (*bottom*).

COURTESY OF THE GUILFORD TOWNSHIP HISTORICAL COLLECTION, PLAINFIELD–GUILFORD TOWNSHIP PUBLIC LIBRARY

"What a few more seasons will do to the ducks." This cartoon by Jay "Ding" Darling, which appeared in *Thirty Years War for Wild Life,* perfectly captures Hornaday's thinking.

COURTESY OF THE JAY N. "DING" DARLING WILDLIFE SOCIETY

5

CAMPFIRES AND CONSERVATION

IN AUGUST 1905, William Hornaday took a much-needed and long-awaited vacation to British Columbia, Canada, with his friend John M. Phillips. "I am about starting West on a hunting trip that I think will do me a lot of good, both physically and mentally," Hornaday wrote to his former collecting companion Chester Jackson. Hornaday enjoyed vacations from work and appreciated their restorative attributes. But he fretted about this one. The old hunter was out of practice: it had been years since he last took up a gun in pursuit of wildlife. He was not a young man anymore, and he knew it. Before the trip, he warned Phillips he was "precipice shy! Having given 'hostages to fortune'" already. Phillips, a mine executive from Pittsburgh, Pennsylvania, harbored no such doubts. An ardent sportsman, and founder of the Lewis and Clark Club, one of the numerous outdoor associations formed in the early twentieth century, he sat on the Pennsylvania Game Commission and played a decisive role in the passage of the Keystone State's pioneering game-protection laws that included mandating hunting licenses, a ban on the sale of game meat, and an alien gun law. When the two men headed west, across the Canadian Rockies, they were entering a veritable hunter's paradise, a relatively untrammeled, although not remote, mountainous region seventy miles north of the border with the United States. Phillips assured Hornaday that the area was "well filled with game, and as yet wholly unspoiled by hunters." Whatever his justifications about hunting mountain goats in the name of science, Hornaday could not resist Madison Grant and Henry Fairfield Osborn's praise of the region and its wildlife.[1]

Dazzled by the rugged landscape, Hornaday instantly concluded that its boosters had not oversold it one bit. Hornaday and Phillips each paid fifty dollars for their nonresident licenses, hired two guides, a scout, a cook, and eight horses, and set into the "unspoiled Rocky Mountain paradise" in search of fish and game. "We are stiff, sore and banged up as never before but severely happy in the knowledge that this was the greatest ever (up here), and we broke no bones either," Hornaday wrote Josephine when he emerged from the mountains. "We had a very strenuous time." They returned with trophies to be mounted, adventures to recall, and a few live animals for the New York Zoological Park, including its first-ever wolverine. Hornaday decided to share his experiences with a larger audience by writing a memoir of his hunting trip.[2]

In June 1906, Hornaday was awarded an honorary doctorate degree from the University of Pittsburgh in Pennsylvania. His brother Calvin jokingly wrote to him as "My Dear Doctor" when sending along a clipping of a press release from a local newspaper. From that point on, Hornaday was addressed as "Doctor," with all the weight that term implied, even if the honor sat uneasily at times with the recipient. "I never was a professor in any institution of learning and therefore always did my best to avoid the bestowal of that title," Hornaday commented in 1920. Yet it was used, with respect by his friends and with a sneer by his enemies who knew he had not earned a baccalaureate degree and doubted his scientific credentials.[3]

After a four-month convalescence leave for a painful ear infection in the summer of 1906, Hornaday resumed his seat as director of the New York Zoological Park and entered the most controversial episode of his life, the affair of Ota Benga, an African pygmy he displayed in the Bronx Zoo. Ota's path from the Dark Continent's bush country to the Bronx was not a direct one. In 1904, Samuel Verner, a missionary sent to Africa to recruit pygmies for the world's fair, had persuaded the 4 foot 11 inch Benga to come to America. Benga had received his introduction to modern, urban, industrial society as a living display in the anthropology exhibit at the St. Louis Fair. Verner was not necessarily a protective guardian; dreams of financial reward from displaying what he considered a species standing somewhere on the evolutionary continuum between apes and humans clouded his morality. At the conclusion of the exposition, Verner had returned with Benga and the other pygmies to Africa. When Verner decided to return to America in 1906, his African companion demanded to join him. "Ota said he wanted to go back to America," Verner wrote in 1916, "and with some misgivings, I permitted him to come." Verner might have overplayed his objections here. He saw profit in having the pygmy at his side during speaking engagements.[4]

Verner and Benga had found themselves in New York City in August 1906 trying to sell some animals to the New York Zoological Society. During the discussions of the sale, Verner noted that he intended to take a trip south to visit relatives, but was unsure of what to do with Benga, who would be unwelcome in the segregated South. At that point, Madison Grant and Hornaday promised Verner that they would provide the African a job and living quarters. "Obey Chief Hornaday," Verner told Benga before departing, "he is our friend."[5]

Considering Benga's past as a "living specimen" and Hornaday's long-standing interest, as expressed to George Bird Grinnell in March 1896, in creating a human anthropology display, it is not surprising that the African's job was to "care" for the chimpanzee at the zoo and that his "quarters" were the monkey house. "A trap was being prepared," wrote Verner's grandson Phillips Verner Bradford and his coauthor Harvey Blume, "made of Darwinism, Barnumism, pure and simple racism." The situation unraveled within a week. On Saturday, September 8, 1906, Hornaday announced to the visitors that they could expect something "new under the sun." This was only the titillating warm-up for next day, the busiest of the week. On Sunday, attendance spiked as throngs of excited New Yorkers, attracted by headlines like "Bushman Shares a Cage with Bronx Park Apes," journeyed to the zoo to see its latest and strangest exhibit. That morning, keepers placed a sign on Ota's cage: composed by Hornaday, it described the African as if he were another exotic animal. On Monday, African American clergy met with one of the country's most notorious racists, Madison Grant, to demand the release of the "living specimen" from his inhumane conditions. Grant demurred, but permitted the African greater freedom in the park.[6]

Despite the opprobrium of the ministers, media outlets like the *New York Times*, Mayor George McClellan, and many of the city's men of science and medicine defended the decision to display Benga in the zoo. Osborn and Grant enjoyed the attention. "The enclosed clippings are excellent," Osborn delightedly told the director. "Ota Benga is certainly making his way successfully as a sensation."[7]

Benga's walks around the park caused almost equal excitement to the patrons and to Benga, who threateningly drew his knife on one occasion. Frustrated, Benga made frequent trips to the director's office and blew his harmonica loudly in Hornaday's ears whenever he got the chance. "Boy had become unmanageable; also dangerous," Hornaday wired to Verner on September 17. "Please come for him at once." Verner removed Benga from the park and released him into the custody of the Colored Orphan Asylum in Brooklyn. Benga left the zoo on September 27, never to return. He did not prosper, however, and his story sadly ended with his suicide seven years later.[8]

Although it lasted but two weeks, Ota Benga's treatment at the zoo has sullied Hornaday's reputation, as it should. It is clear to us a century later that this act of cruelty was wrong. Nonetheless, responsibility for this deplorable event does not rest on Hornaday's shoulders alone. Madison Grant and Samuel Verner share responsibility for bringing Benga to the Bronx. More distressing, men of science (Henry Fairfield Osborn, for example) raised no objection to using Benga as an object lesson to promote their conceptions of human evolution with its implications of racial hierarchy.[9]

As summer turned to fall in 1906 in the wake of the Ota Benga controversy, Hornaday completed the memoir of his hunt in British Columbia, *Camp-Fires in the Canadian Rockies*. Charles Scribner's Sons released it that December. It was a classic cry for a return to nature written expressly for the "tired businessman, the overworked professional man, and the sleepless newspaperman." Hornaday urged this class to "get next to the soul of Nature" as an antidote to the toxins of modern life. While Hornaday loved his job as director of the New York Zoological Park, the demands of the patrons, city officials, and his blue-blood bosses in a city of millions of people speaking dozens of languages crowded into a space that was entirely too small took their toll even on his proven nerves. He had had to escape the pressures of work, and he aspired to convince his readers to emulate his refreshing experience.[10]

Camp-Fires in the Canadian Rockies also contained an important message on the conservation ethic aimed specifically at hunters and conservationists. Throughout the book, Hornaday emphasized that the pursuit of game paled in importance compared to a genuine communion with nature. As in his book *The American Natural History*, he called on his readers to put down their guns in favor of the camera. The idea of collecting photographic trophies instead of mounted ones had begun in late-nineteenth-century sporting magazines like *Forest & Stream, Recreation,* and *Shields' Magazine.* "The camera hunter is rapidly crowding the man with the gun to the rear," George Shields claimed in 1906. But it was a tough sell because the cameras of the day were large, bulky objects that did not lend themselves to being easily carried deep into the forest to take pictures of small birds and moving mammals through dense foliage. And poor resolution often provided disappointing results. Plus, there was something of a stigma attached to photography as being unmasculine compared to the primal act of hunting. Hornaday attempted to address these concerns in *Camp-Fires in the Canadian Rockies* by vividly describing Phillips's harrowing adventures with the camera. The Pennsylvanian climbed trees, hung over cliffs, and got much closer to dangerous animals with his camera than he would have with a gun.[11]

Hornaday was deeply concerned about the wildlife he had left behind in British Columbia, however, and both he and Phillips championed a wildlife refuge for the area they had hunted. The local Fernie Game Protection Association acceded to the need, but the supportive local game warden, A. Bryan Williams, warned Hornaday that with upcoming local elections, "it is about out of the question to attempt anything" in the immediate future. Yet, like the railroads of the United States, the Canadian Railway, which owned the land, wanted to tap into the tourist trade a preserve could generate. With the local game commissioner, sportsmen, and landowners in favor of a preserve, Hornaday had cause for optimism. "At first Mr. Phillips and I fondly—and foolishly—imagined that if we would make a flying start for an Elk Mountain game preserve, everybody would help," he wrote in a handwritten note for his scrapbook.[12]

The local sportsmen, however, conceived of a very different type of wildlife preserve, and offered a plan that Hornaday saw as "a hostile counter proposal" and a "wild new scheme." To some extent, the difference between Hornaday's vision and those of the Fernie sportsmen echoed a classic conservation story of outsiders attempting to determine resource usage for a local population. Throughout the late nineteenth and early twentieth centuries, elite urban sportsmen often saw the local use of game as wasteful. They criticized the motives of local hunters and condemned any killing of game animals for the marketplace, even if the exchange was among neighbors. In the case of a game preserve in British Columbia, the Fernie hunters wanted to emphasize the animals that mattered most to their economy, deer and elk, because they were an important dietary supplement, as for many rural areas in North America. Like many conservationists, Hornaday had no qualm about telling the local populations what policies were best, and placed priority on what wealthy sportsmen might travel to western Canada to shoot the mountain sheep and goats, which were fast disappearing from North America and largely unavailable in the United States. Game Warden Williams supported the Fernie proposal. He argued that despite the needs of the continent, British Columbia really did need to protect its deer, and that there were more than enough sheep and goats to meet local needs. But to Hornaday's mind, any park that did not include mountain goat was "nothing less than an outrage."[13]

Hornaday overcame the local opposition by cultivating important allies within the province. He always understood the power of the press and was a favorite of newspaper publishers for his flamboyant, attention-getting style. He won over the *Victoria Daily Colonist* newspaper, the local mayor, several important local sportsmen outside of the Fernie Association, and the provincial premier. "But for that friendliness," Hornaday wrote of Premier McBride's support, "we could not have carried on." More importantly, he appealed to the pocketbook of the landowners, the Canadian Railway, by taking a page out of the playbook that had helped to create Yellowstone National Park in 1872. After he convinced the railway that the land his preserve would protect was rocky, steep, and devoid of mineral or timber assets—essentially economic wasteland—he argued that the wealthy sportsmen who would come in from the United States would provide more revenue from traffic and lodging than could otherwise be gained from the railway's useless real estate. On the other hand, the Fernie proposal would close or restrict coal-mining operations.[14]

In February 1908, Hornaday secured the support of the North American Fish and Game Protective Association. Two weeks later, H. W. Herchmer, president of the Fernie Game Protective Association, wrote to John Phillips to say that his organization was willing to discuss a compromise. "The enemy runs up the white flag!" Hornaday triumphantly wrote in the margin of his scrapbook. The government of British Columbia created a commission to pick from the rival plans and dispatched Williams to make an official report after a detailed survey. The panel followed his recommendation and established the Elk River Game Preserve on November 15, 1908, with almost all the features exactly as Hornaday had first proposed, including the primary focus of protection on goats and sheep.[15]

While the debate over the Elk River preserve was raging in British Columbia, Hornaday was relishing another adventurous hunting expedition, this time to the Sonoran Desert along the border between the Arizona Territory and Mexico. Dr. Daniel Trembley, director of the New York Botanical Garden, who was collecting desert specimens for the Carnegie Museum in Pittsburgh, had invited Hornaday along for the adventure. There was an air of romance about this southwestern expedition that had been absent from his adventure north of the forty-ninth parallel. The mysterious desert region was terra incognito to whites, a blank spot on the map where even the source of the Sonora River could not be located. "To the reading and thinking world it was totally unknown," Hornaday wrote. As always when he transited west, he made arrangements to visit with family and friends along the way, even if only briefly.[16]

Hornaday left his household at a delicate moment. His daughter, Helen, had recently become engaged to George Fielding, a Kansas-born executive at Westinghouse. While the couple would not marry until her father returned, the Hornaday home on Decatur Avenue was abuzz with planning and preparations. The three members of the Hornaday family were a very close, tightly knit unit, and Helen's marriage threatened to substantially alter their personal lives. "What her mother will do without her,—heaven only knows, for *I* don't!" Hornaday wrote to his half brother David. But the marriage proved to be less traumatic than Hornaday at first supposed. Helen rarely strayed far from her parents, and Hornaday had great affection for his son-in-law, who, in turn, had ample respect for him.[17]

Hornaday also left for the Sonoran Desert on the heels of an imbroglio at the zoo. It was characteristic of Hornaday to attach great significance to small and even trivial events, and in April 1907 he had gotten into a heated altercation with Dr. Bynum of the American Museum of Natural History over the routine exchange of a pair of elk antlers. Because Henry Fairfield Osborn was chairman of the New York Zoological Society's Executive Committee and president of the American Museum of Natural History, the two organizations worked in close concert. They shared technical services, specimens, and supplies. Animals that died at the New York Zoological Park often found immortality as displays at the American Museum of Natural History. Madison Grant cautioned Hornaday against personalizing business transactions related to the society, and to Osborn he expressed concern over the director's prickly nature and unprofessional attitude toward colleagues. "I have told Hornaday many times that Bynum feels very kindly towards him," Grant reported to Osborn. "These imaginary grievances of his are so absurd that it is scarcely worth while protesting any more about them." But the matter of the elk antlers would not go away. "What on earth will we do? Am getting a little tired of these endless broils," Grant wrote in exasperation to Osborn two weeks later. Grant and Osborn felt Hornaday's lapses of dignity and tact embarrassed the society, but certainly class also played a role in their relationship with Hornaday. They were wealthy, patrician New Yorkers with Ivy League educations and cosmopolitan views who probably considered their director something of an Iowa hayseed with parochial tastes. On the other hand, Hornaday was a self-made man with all the sensitivities of the type, and while he liked mingling with the social elite on one level, he resented their snobbery on another. Class anxiety might have also soured his relationships with other patrician conservationists like John Bird Burnham and George Bird Grinnell and contributed to his sense of being an outsider.[18]

The Sonoran expedition set out in early September 1907. In addition to Hornaday and Daniel Trembley, D. T. MacDougal of the Tucson Desert Botanical Laboratory, Hornaday's friend John M. Phillips, Jefferson Milton of the Immigration Service, and Godfrey Sykes, who acted as cartographer, joined the expedition. This botanical collecting expedition was big news in Arizona. The *Tucson Citizen* covered their progress, and papers in New York, Pittsburgh, Washington, D.C., and even London carried the stories. The team set out from Tucson in a wagon train destined for the Papago Tanks, a natural oasis. After a week there, they headed south toward another oasis called Tule Tanks. Then they planned to go literally off the map into uncharted desert wilderness. As they filled in maps, they added their own names to the geographic features, saving the largest, grandest one for their patron, Andrew Carnegie.[19]

To President Roosevelt, Hornaday described the Sonoran as "a real desert *botanical garden*." Amid the craters, creosote bushes, and mesquite trees, there were singular and colorful flowers and towering sixty-foot-tall cacti. Hornaday felt that thousands of other Americans thought of deserts as he had before his trip, as stark, barren, and devoid of life. To dispel this myth, he took pen to paper to describe his adventure. *Camp-Fires on Desert and Lava* came out in October 1908. "One can hardly read Mr. Hornaday's book without wishing that he could go and spend a holiday in that same arboreal desert," commented one reviewer.[20]

Camp-Fires on Desert and Lava contained the same call for sportsman's ethics as had *Camp-Fires in the Canadian Rockies.* Hornaday urged hunters to kill as little as possible, shoot only for the table and scientific purposes, and realize that the true enjoyment of a hunting trip was the experience of nature, not killing game; again, he recommended the camera over the gun. Everyone in the team had agreed that there would be "no unnecessary shooting of any kind." But *Camp-Fires on Desert and Lava* differed from its predecessor in one important respect. While Hornaday had had little to say about the local Anglo population in Canada, he openly detested the Native Americans and Mexicans of the Southwest. In racial comments reminiscent of those he expressed in *Two Years in the Jungle,* Hornaday wrote of their laziness and dirtiness. The best he could say about Mexicans was that their "language does not grate on the ear like the filing of a ripsaw." At worst, the expedition was highly suspicious of their Mexican guides and watched them constantly in fear of thievery. They were even less friendly toward the Papago Indians. When running low on water, Hornaday and company felt totally justified in robbing two Papago at gunpoint for their supply. "If there must be another Indian outrage, why there was no better place for it than in that silent plain," he wrote with thinly veiled machismo, "where graves might be had for the asking."[21]

Hornaday's contempt for other ethnic groups sometimes seemed to know no bounds. He wrote disparagingly not only about the people he encountered in the southwestern desert, but also about the people he left behind in the urban Northeast. He detested the "new" immigrants from southern and eastern Europe, and was glad to leave them and their "Bowery English," as he dubbed it, far behind. The desert, and by extension all wilderness, was a place of "boundless space" where the spirit could roam freely, and where there was an "absence of dust, filth, waste paper, polluted streams, dirty humanity, and many other things that wear on Life in a great city," immigrant masses included. A year after he returned from the southwestern deserts, Hornaday began his "rubbish war" against guests who dared litter in his zoo. He was dumbfounded that people would sit within sight of a garbage can and yet throw their trash on the walkways and grass. It greatly annoyed him that the New York Zoological Society spent as much as three thousand dollars a year to clean up after such blatant slobs. He also feared that trash dumping, which he exclusively blamed on the immigrants, would lead to more serious crime at the zoo. Throughout his tenure as director of the zoo, he constantly complained of patrons who stole from the zoo and either goaded or taunted the animals with sticks, bottles, and even lit cigarettes. By strictly enforcing the rules regarding rubbish, he hoped to prevent more serious transgressions. He frequently chided the local police station for failing to provide enough officers and plainclothes detectives to patrol his zoo. However, Hornaday never felt like he won the rubbish war, which dragged on throughout the following two decades.[22]

In early 1908, Hornaday introduced a proposal to establish another federal buffalo preserve. His plan followed the successful model of federal-private cooperation first implemented with the Wichita Forest and Game Preserve. But instead of working through his employers this time, Hornaday acted as an agent for the recently formed American Bison Society (ABS) to provide the animals. When he introduced the plan, he was serving as the president of the ABS, but he embraced the role with uncharacteristic caution. It had not been his idea, but that of Ernest Harold Baynes, an amateur naturalist. T. Gilbert Pearson had attended a Baynes lecture in Boston entitled "The American Buffalo, A Plea for Its Preservation" and referred him to Hornaday. At first Hornaday gave Baynes the cold shoulder. The director did not want to create an organization that appeared to be competing with the New York Zoological Society. But he relented by the end of the year, and the American Bison Society was formed on a cold December night in the Lion House at the New York Zoological Park. It existed mostly on paper, and failed in its first effort to persuade New York State to establish a preserve in the Adirondacks in 1907.[23]

In March 1908, Senator Joseph Dixon, Republican of Montana, sponsored a bill to create a twenty-square-mile buffalo refuge in the Flathead Indian Reservation along the lines proposed by Morton Elrod, a University of Wyoming biology professor and ABS member. Later that month, Hornaday traveled to Washington, D.C., and secured the blessing of President Roosevelt during a private meeting at the White House. "Hornaday," the president told him, "whenever you come to a point where you really NEED me, and can't get forward without my help—you come and see me!" The Senate passed the measure and the thirty-thousand-dollar appropriation quickly, but the bill stalled in the House. "Our task with the House is serious!" Hornaday wrote to his friend Charles Davenport in mid-April, but he called in the Rough Rider for reinforcements and the bill passed. President Roosevelt signed it into law in mid-May. The Kansas Republican congressman Charles Scott, chairman of the Agricultural Committee, reminded Hornaday, "it is now up to you to corral the Buffalo!" That, however, was easier said than done.[24]

Unlike the campaign to create the Wichita preserve, Hornaday did not have buffalo on hand for the Flathead preserve. But this did not overly concern him. He thought he could persuade Michael Pablo, a Flathead Indian who had already sold his herd of 350 purebred buffalo to Canada because the federal government opened the reservation commons on which he kept his herd, to set aside thirty to forty of the animals for the ABS. "It will be a great disappointment to me if he lets the whole lot go to the Canadians," Hornaday wrote to a fellow ABS member in early April 1908. However, Pablo was angered at being forced to sell his animals by the federal government's opening of the reserve and further aggravated by Hornaday's ridiculous appeals to patriotism. He even threatened to "shoot them first before he would sell them [to the American Bison Society]." Hornaday further erred in offering $200 per head, even though the Canadians had paid $350, the same price as prime beef. Charles Aubrey, a white Indian agent with a Flathead wife, described Hornaday's tone-deaf approach to George Bird Grinnell as nothing less than blatant racial hostility and "an unpardonable sin." In Hornaday's defense, it was less a calculated insult to a savvy businessman and more the result of ignorance of the details of the Canadian offer. Either way, all 350 animals went north to the Banff Reserve near the Great Slave Lake, at the northernmost point of the buffalo's natural range. Hornaday and the ABS would have to find their animals elsewhere.[25]

Even though the Pablo herd was off the market, Hornaday set about the task of raising the ten thousand dollars he gauged to be the necessary amount to purchase buffalo for the Flathead preserve. "Dr. Hornaday's patriotic labor in collecting the money required for the first wild herd will not be unrequited," the *New York Times* editorialized. But it was demanding, if patriotic, labor. As he had done during the Elk River preserve campaign, Hornaday solicited personal contributions, with the vast bulk of individual donations amounting to less than five dollars each. It was exhausting work, and Hornaday vowed never again to raise money by subscriptions. "It nearly killed me," he wrote in 1911.[26]

Hornaday ultimately found a herd available from the estate of the late C. E. Conrad, an industrialist who owned a moderate number of animals not far from the Flathead reservation. After some negotiations with the estate, the ABS purchased thirty-four animals at $275 a head for a total of $9,350. Within ten years, this small herd grew to two hundred head. By 1924, the herd at the Flathead Buffalo Refuge numbered 560. For Hornaday, the work of reviving the buffalo was complete. "I regard the American bison species," he declared in 1913, "as now reasonably secure against extermination." Later in life he convinced himself that his efforts to protect the buffalo atoned for his killing from the small surviving band in Montana in 1886.[27]

Hornaday was correct in his assessment that the buffalo had been saved from extinction, although it bears pointing out that he alone did not "save" the buffalo. George Bird Grinnell played a decisive role in the 1890s preservation of the small remnant of wild animals in Yellowstone National Park; John F. Lacey guided several key bills through Congress; and President Theodore Roosevelt signed most of them. Nor should saving the buffalo be considered an achievement of white sportsmen and politicians. There would be no animals to stock the preserves if Native Americans had not built up private herds when the wild ones were hunted out of existence by white market hunters. For too long the Native Americans have been denied their important contribution to the preservation of the animal so important to their culture. In another sense, however, the buffalo was lost to history. The enormous wild, free-roaming herds are gone forever, replaced by their tamed heirs who are confined to preserves and parks, and killed if they stray across the unnatural boundary.[28]

In early 1907, before he had departed on his expedition to the Sonora Desert, Hornaday had received an urgent appeal from Henry Wood

Elliott begging him to "do something to save those fur seals!" Elliott, a self-described naturalist with no formal training, was referring to the seals of the Pribiloff Islands of Alaska. Negotiations between the United States and Great Britain to reduce the number of seals killed at sea by Canadians had broken down, and Elliott had turned to Hornaday as his last hope.[29]

Hornaday could not refuse such a desperate plea to help a species that he believed stood on the precipice of extinction. Although he was a relative newcomer to a tumultuous controversy already four decades old, he was by no means uninformed. He had been collecting facts about the conditions of the seals since the early 1890s, when a dispute over the seals greatly strained relations between Great Britain and the United States, which he catalogued in *The American Natural History*. The conventional wisdom of scientists familiar with the animal, as well as the federal Fur Seal Service, maintained that pelagic sealing—a wasteful practice in which females and juveniles were slain along with males at sea—threatened the herd with extinction. They did not question managed seal harvesting on land. In fact, they supported culling the herd on the grounds that an excessive population of fighting males trampled the pups and hence threatened the viability of the herd. Elliott and Hornaday challenged this consensus. As was the case with other species, Hornaday earnestly believed that any connection between a species and the market guaranteed its extinction. In responding affirmatively to Elliott's appeal for help, Hornaday entered into one of the bitterest conservationist efforts of his career. His preservationist ideology ultimately triumphed over the utilitarianism of the scientists. In some respects, the battle over the seals mirrored the similar confrontation between the preservationist John Muir and the utilitarian Gifford Pinchot over the flooding and damming of the Hetch-Hetchy Valley in Yosemite National Park to supply water for San Francisco. In both cases, earnest men clashed over priorities.

After some tentative moves, Hornaday decided to wait until the Senate opened discussions in 1910 on the subject of renewing the lease for the company with the monopoly license to harvest seals on land. Then Hornaday would propose a ten-year closed season for land killing in conjunction with diplomatic efforts to end pelagic sealing. This position stood in marked contrast to the conventional wisdom of the scientific community that land sealing did not represent an existential threat to the herd. Hornaday's connection with Senator Joseph

Dixon, chairman of the Committee on the Conservation of Natural Resources, would provide him with not only the opportunity to offer legislation, but great liberty to dominate the hearings. Nevertheless, Hornaday hoped to stand apart from the fray personally as much as he could. "I propose to go in as an interested bystander, friendly to all parties, backed by a large following of people, who are also non-partisan, and demand in the name of the American people that the present intolerable situation be brought to an end in a businesslike and also speedy manner," he wrote to Elliott on November 1, 1909.[30]

Whatever his intentions, Hornaday did not remain nonpartisan for long, and it was impossible for him to remain a "bystander" or "friendly to all parties." He did, however, put a clamp on Elliott's controversial tongue and asked that old history remain in the past. "Therefore, old fellow, don't explode any bombs now," he pleaded to Elliott four days later. "I am going to urge Senator Dixon not to go any further than necessary into the history of the fur-seal industry," Hornaday also wrote to Elliott, "because the moment we begin to dig up the past, the friends of the fur seal will get into trouble with its enemies." But this, likewise, proved impossible in the charged atmosphere resulting from twenty years of ad hominem attacks, wounded feelings, and even allegations of treason.[31]

In February 1910, Hornaday testified before Dixon's friendly Senate Committee on Natural Resources, where he recommended the elimination of the lease system, implementation of a ten-year closed season on land sealing, and a treaty to eliminate pelagic sealing. His appearance had the desired effect, and it looked as though a closed-season bill would sail through Congress. Secretary of Commerce and Labor Charles Nagel, whose department had had oversight of the herd since 1903, persuaded enough senators to reword the bill to give him the discretionary power to proclaim a closed season, but not explicitly require him to do so. Senator Dixon did not see this as a major shift in the meaning of the legislation because he believed there was universal acceptance that a closed season was in order. Nine days after the bill passed, however, Nagel ordered George Bowers of the U.S. Seal Service to resume the seal harvest.[32]

Hornaday fumed at Nagel's decision. "This action is an act of war," he dramatically wrote Henry Wood Elliott. Hornaday later testified before the Senate that the decision was a "mighty shock." He voiced his outrage directly to Secretary Nagel: "Did the President . . . or the

United States Senate, intend for one moment that you should go right on in the bloody killing business without a halt? No! A thousand times, no, and you know it!" Nagel curtly responded with indignation of his own. He reminded Hornaday that he had had his chance to persuade Congress to legislate a closed season and had failed to do so. Empowered by Congress with full authority to make an informed decision regarding the fur seal, and having made it with the input of his acknowledged panel of experts, Nagel "considered the question closed." "Now Mr. Hornaday," Nagel added, "you have considerable responsibilities in your official employment, and I shall endeavor not to molest you. I hope that you will accord me the same privilege in my capacity."[33]

In fact, the matter was not closed as Nagel had wished; indeed, it was only just beginning. Accepting the secretary's challenge, Hornaday responded, "your implied threat to me if I pursue this matter any further is of no effect anywhere." He added with more bravado, "you are welcome to 'molest' me if you can." Every time Nagel replied to him, Hornaday fired off another letter. After having baited Nagel—who walked into the trap easily enough—Hornaday published the correspondence in a pamphlet entitled *A Square Deal for the Fur Seal* in early July. His strategy of comity, less than a year old, lay in ruins, but he did not lament it. "Regarding our fur-seal war," he wrote to Henry Fairfield Osborn, "I think we have routed the enemy, horse, foot, and dragoons."[34]

Early in 1911, Senator Dixon renewed his hearings on the grounds that Secretary Nagel was not following the intent of the bill passed the prior year. When Hornaday brought up the issue of the gentleman's agreement at the hearings on February 4, 1911, members of the committee confirmed their belief that Nagel had assented to halt harvesting. "I never believed in permitting the killing upon land at all," Senator Weldon Heyburn, an Idaho Republican, commented. Hornaday continued to assert that both land and pelagic sealing had to end for the seal herd to recuperate. "The killing of seals on land and the killing of seals at sea by the pelagic sealers has gone on until the fur-seal life has been drained to the very dregs," he said in congressional testimony. Henry Wood Elliott also took advantage of the prominent platform Dixon's hearings afforded by airing sensational allegations against the U.S. Seal Service. Citing figures from the *Fur Trade Review*, a London publication, Elliott charged that the U.S. government killed an astounding 7,700 yearlings in July 1910, violating its own rules and threatening the viability of the herd. Seal Service officials challenged Elliott's evidence, countering that no correlation existed between the weight of a dressed seal skin in London and the actual age of the living seal in the Pribiloff rookeries. Under oath, Walter Lembkey, the seal agent on the Pribiloff Islands and the man in charge of the government's harvest, emphatically refuted Elliott's allegations.[35]

The House, now controlled by Democrats after the election of 1910, attacked Nagel more vociferously. Representative John Rothermel of Pennsylvania, chairman of the Committee on the Expenditures of the Commerce and Labor Department, drilled in on Elliott's allegation that the government had killed yearling seals. Elliott relished his role as the committee's chief seal expert, using his power to engage in a war of personal vindication, dredging up events thirty years old in what at times resembled a witch hunt against the Seal Service. With a taxpayer-funded office space, stenographer, and franking privileges, he dragged the hearings out for three years. Hornaday was cautious about the hearings at first. "A prolonged investigation may have the effect of wearing everybody out," Hornaday warned Elliott at the end of October 1910, "the way the Ballinger investigation did." But his attitude changed after the Democrats gained power. As the hearings progressed, he saw a practical value to Elliott's exposure of the alleged misdeeds of the Seal Service. To Edmund Seymour, he wrote in June 1911, "I really believe that the investigations now going on will clear the atmosphere, and do the cause of the fur seal a great deal of permanent good." But it was only the beginning of the storm.[36]

While Elliott attacked the Seal Service in Washington in late 1910 and early 1911, Hornaday prepared to lobby for new legislation in New York State that would ban the sale of game meat. "I have become strongly possessed of the idea that now is the time to strike for some radical measure in the line of wildlife protection," he wrote to Madison Grant. It was not a novel idea. Nineteen other states had restrictions on game-meat sales to varying degrees, and George Bird Grinnell had proposed such a law for New York State in a *Forest & Stream* editorial in 1894. In *Thirty Years War for Wild Life,* a memoir shaped by two decades of disputes with Grinnell, Hornaday had taken pleasure in mocking the 1894 editorial by writing, "the editor of a New York magazine had hung up as a shibboleth the words 'Stop the sale of game,' but his labors in that vineyard had gone only to that limit." Now, seventeen years later, Hornaday dedicated himself to doggedly laboring in the vineyard and to not leaving until he reaped a harvest.[37]

A game-meat ban in New York was not solely a matter of state interest; it was a national conservation issue. Hundreds of thousands of game animals in the southern and eastern United States were killed and shipped to hotels, market stalls, and restaurants in the nation's largest city. Less than a decade before Hornaday took up the cause of game-meat sales, just one New York City cold storage facility was found to contain 42,759 birds. August Silz, a Brooklyn game-meat dealer, claimed to sell more than a million game animals each year. Closing the New York game market would save dozens of species and millions of individual animals east of the Mississippi River. Accordingly, Hornaday's attorney, Lawrence Trowbridge, drew up a stern bill that differed from existing state laws by banning the sale of game meat year-round, without exception, and applying stiff fines for violations.[38]

Using the Progressive Era language of the day, Hornaday tied the sale of game and the destruction of the common national wildlife heritage to the greed of the plutocratic few, even if the numbers suggested this small group could not possibly have consumed all the meat sold. "It is the commercial spirit seeking special privileges, at the expense of the property of the people of the whole state," he wrote in an editorial for the *Daily North Side News*. Grinnell had noted in 1894 how unnecessary game meat was for the food supply, and Hornaday recycled this theme in his own sardonic style when he declared, "The people who pay $3 for a roast wild duck do not know what hunger is, save by hearsay." In the *New York Tribune*, he claimed that the unrestricted sale of game meat undermined the entire hunting regulation system because half of all the game killed out of season was killed for the purpose of selling it. Why bother with licenses or closed seasons if the market provided such an enormous incentive to break the law? By advancing these justifications, Hornaday presented the Bayne bill, as the New York legislation was known (after state senator Howard K. Bayne of Staten Island), as a universal solution that would restore respect for law and order, bolster all the other game regulations, protect the public property against selfish depredation, and save millions of game animals across the United States. And all this could be done at practically no cost to the taxpayers.[39]

Having gained the support of all the important conservation organizations of the state, Hornaday launched an aggressive lobbying effort that included pamphlets, newspaper editorials, interviews with members of the press, and a mail campaign that sent more than thirty thousand supportive letters flowing into the mailboxes of state legislators. This effort created some tension with his employers at the New York Zoological Society, as he received both praise and warnings. "I have been reading your *Wild Life Call* with care," Henry Fairfield Osborn wrote of one of Hornaday's pamphlets, "and I consider it a very strong and convincing document. I was glad to see a stirring editorial in the *Times* this morning. The work you are doing in this line is magnificent." Yet, Madison Grant did not lighten his grip on the leash and sternly warned the director not to write any "yellow pamphlets" on society stationery.[40]

On March 25, 1911, the Triangle Shirtwaist fire killed 146 garment workers in New York City. This tragedy diverted the attention of state lawmakers away from the Bayne bill and every other piece of legislation as they established a committee to investigate the appalling conditions that had led to the fire. Four days later, Hornaday, Madison Grant, and Charles "Buffalo" Jones led a junket to Albany to testify before a committee and directly lobby lawmakers. Over the next two weeks, Hornaday learned that his strong game-meat ban did not have the support it needed among the lawmakers, and that the game-meat dealers were open to compromise.

On April 7, Hornaday assented to a revision that would allow licensed game dealers to sell farm-raised deer, ducks, pheasants, and rabbits. In an interview with reporters, Hornaday announced that the issue was "so full of puzzles" that it required "future consideration and action," code words for compromise. But he felt that he had already won a victory by getting the game dealers to the table. "Wonder of Wonders! The game dealers and their counsel have accepted the principle that no wild game shall be shot for the market, or sold for food!" he exultantly wrote to one correspondent. Two weeks later, the game dealers demanded another revision, this time to add three foreign bird species to the list of allowable species. This presented a dilemma for Hornaday because Theodore Palmer at the Bureau of Biological Survey warned him that it would be unwise to permit those species to be sold, as it could potentially complicate conservation efforts in other states. Hornaday relented, however, and advised the bill's sponsor and namesake, Senator Howard K. Bayne, to rewrite the legislation once more. "We can not afford to lose the main features of the Bayne bill on account of that!" he exclaimed. Nevertheless, this latest concession did not assure passage, and Hornaday despaired that the opposition was either chipping away his prized reform one animal at a time, or dragging out negotiations until the legislative session expired. "We have done everything that mortal men could do to induce the legislature to do the right thing," he wrote. "For myself," he wrote to Senator Bayne in an effort to draw a definitive line in the negotiations, "I am unalterably opposed to further compromises with the enemy." Bayne, however, agreed to additional amendments. "I had reason to believe that the Bill would not have gone through if I had refused to accept the changes," he explained to Hornaday, "and a defeat at that time and in that way might have been disastrous beyond repair." The bill passed both houses with only a single negative vote cast. New York governor John Dix gladly signed it into law in early June, an action that earned him Hornaday's approbation as the "grand conservationist governor."[41]

In one very important aspect, Hornaday took away the wrong lesson from his Bayne bill experience. "My great no-sale-of-game bill soon will become a law!" he excitedly wrote to former president Theodore Roosevelt when it passed the legislature. "When I had it drawn and introduced, some of my friends said I was crazy!" This might have been the case, but it was a modified bill and not the original one that passed. His enemies emphasized this fact in later years, largely because Hornaday was prone to boast of his accomplishment. He should have learned the value of compromise. Instead, he focused on how few people thought he had a chance to get his measure through the legislature, and how the lesson he learned was to ignore the advice of others by taking the boldest stance possible. In future campaigns, he was less willing to negotiate and compromise, and more stubborn in adhering to his original ideas.[42]

The Bayne bill received universal praise from wildlife conservationists. Even at the height of the imbroglio over the acceptance of money from gun manufacturers, the NAAS fully supported the measure in the strongest terms. In 1912, California and Massachusetts followed New York's example by enacting strong bans on the sale of game meat. The historian Frank Graham wrote that by closing down the largest market in the nation, the Bayne bill "virtually ended the large-scale traffic in native wild game in the United States." Certainly it is true that the Bayne bill greatly reduced the amount of game meat marketed across state lines, but there was one caveat. As the historian

Louis Warren has demonstrated, state bans on the sale of game meat had little impact in rural areas where local residents traded openly among themselves with little legal consequence. This, however, was not the market Hornaday targeted. By closing off the largest city in the nation to the sale of game meat, millions of wild animals would no longer fall prey to the market hunter.[43]

Throughout the campaign for the Bayne bill, Hornaday feverishly worked to secure enough funds to pay for basic lobbying expenses. "It was a case of adequate funds or no fight," as he wrote in *Thirty Years War for Wild Life*. The costs of lawyers, stationery, postage, travel, and lodging far exceeded what he could afford to pay out of his own pocket. The New York Zoological Society had little cash to budget for Hornaday's campaign, and the Campfire Club, a group Hornaday had a hand in founding, put whatever money they could allocate toward wildlife protection into the fur-seal fight. Another alternative would have been to finance each campaign separately through subscriptions, but after the grueling experience of the Flathead Bison Preserve two years before, Hornaday had sworn off that method.[44]

Instead, he opted to raise large sums of money from selected, proven donors for what he called the discretionary fund, which he could spend in any fashion he saw fit. He considered one thousand dollars the amount needed to finance the Bayne bill, but was pleasantly surprised when his wealthy patrons contributed more than five thousand dollars. The donors included a mixture of New York City elites, philanthropic business leaders, conservation organizations like the Boone and Crockett Club, and even three small gun-manufacturing firms. He had to assure both Grant and Osborn that he would not unduly pressure important donors to the New York Zoological Society. After the Bayne bill passed, Hornaday decided to extend the concept with the Permanent Wildlife Protection Fund, capitalized at one hundred thousand dollars. He was *"too tired!"* to "raise any more money by small subscriptions, year by year," he wrote to Henry Clay Frick in 1913 when soliciting funds for the Permanent Wildlife Protection Fund. "I want to spend my remaining time and strength in *directing* campaigns."[45]

The Permanent Wildlife Protection Fund allowed Hornaday greater independence from what he considered the stifling restraints of the New York Zoological Society. Grant and Osborn's constant efforts to limit his conservation activities greatly frustrated Hornaday.

"Then I vowed that I never again would fight under the banner of the New York Zoological Society," he wrote in March 1912 after one such incident. His Permanent Wildlife Protection Fund provided him with a new flag to carry into battle, one he shaped exclusively to fit his own agenda. Although he had to report the fund's expenses to its donors, he never had to ask their permission for expenditures. Despite the establishment of the fund, Hornaday was never completely free of his bosses' constraints, and his finely tuned sense of independence would bristle indignantly on a few more occasions before his retirement in 1926. There were several reasons why he accepted this arrangement. First, he respected Grant and Osborn. He especially valued Grant's opinion, and considered him the best spokesman for the sportsmen. Second, he loved his job and could not imagine himself as anything other than the director of the largest zoo in the world. He might test the boundaries, but he never consciously jeopardized his position. Third, money played a role as well. Having never overcome his financial insecurity, he needed the salary. His books provided a nice supplement to his income, but nothing he could live off. He consistently made between $300 and $600 annually between 1905 and 1919, and from $400 to $1,100 in the 1920s in royalties. In 1923, the year of his highest income as a writer, he earned $2,489.46, a sum that amounted to one-quarter of his salary as director.[46]

6

OUR VANISHING WILDLIFE

IN MARCH 1911, just after William Temple Hornaday returned from his politicking in Albany against the sale of game meat, he met with H. S. Leonard, a vice president with the Winchester Repeating Arms Company. Although Hornaday subsequently produced several different versions of the meeting with varying details, they all agreed on an essential point: Winchester offered Hornaday at least ten thousand dollars a year for wildlife-protection efforts on the condition that he abandon his campaign against the automatic and pump shotguns. Since Browning had introduced the first weapons of this type around the turn of the century, Hornaday had lobbied (largely through the New York Zoological Society's agent, George Shields) state governments to limit the use of repeating shotguns. Echoing a theme from "The Extermination of the American Bison," Hornaday believed this technological improvement provided the gunners with too great an advantage over game, and threatened all bird species with extinction. Only the dreaded market hunters and game hogs would use such a weapon, he argued. During his meeting with Leonard, he urged Remington and Winchester to commit to limit the number of shells their pump and automatic guns could hold to a maximum of two. "This Mr. Leonard positively declined to consider," Hornaday reported to A. H. Fox, the owner of a small shotgun manufacturing firm. "I was greatly disappointed."[1]

Hornaday knew Leonard would try to entice another party to accept their money, and he was equally convinced that it would be a very bad thing for conservation if someone accepted the cash on the gun maker's terms. Such agreements threatened to channel money for wildlife-protective measures toward those who also served the profit margins of the companies that produced the weapons and ammunition used to shoot birds and mammals. "So long as they maintain their present attitude, and manufacture the automatic and 'pump' guns unchanged, it is, in my opinion, utterly impossible for any real friend of wild-life either to work with them, or to accept from them any money contributions whatever," Hornaday added in his letter to Fox. Winchester's proposal and Hornaday's antagonistic response was merely the opening gale in a much larger storm that would shake the conservation movement for decades.[2]

The Winchester Repeating Arms Company did not abandon their effort to influence wildlife protection after Hornaday turned them down. After turning to the New York conservationist Andrew Meloy, who denied their overtures for the very same reasons Hornaday voiced, the gun makers approached T. Gilbert Pearson of the National Association of Audubon Societies. Hornaday had long considered Pearson too vacillating and weak-kneed to lead such an important bird-protection organization. Pearson's chief liability, in Hornaday's mind, was that he failed to follow the exact polices laid out by William Dutcher, the former Audubon president who had suffered a debilitating stroke in 1910. Not only was Dutcher a close friend and supporter of Hornaday, but in 1906 he had put Audubon on record against the use of the deadly automatic and pump shotguns. "This society is working for the preservation of the wild birds and game of North America," Dutcher wrote to Hornaday in 1906, "and it sincerely should not stultify itself by advocating the use of one of the most potent means of destruction that has ever been devised." Hornaday interpreted Pearson's failure to follow through on Dutcher's pledge as nothing less than the betrayal of the cause and his martyred friend. When Hornaday learned in May that Leonard was working with Pearson, he waited patiently, like the skilled hunter that he was, for the most opportune shot. His intent was not to destroy the Audubon Society, but to shame its leadership and arouse its members to take a more active role in policy decisions. The moment arrived on June 2, 1911, when the Audubon Board of Directors voted five to three to accept an annual subscription of twenty-five thousand dollars from the nation's leading gun-manufacturing firms.[3]

Although the meeting was supposed to be secret, the results

leaked to the *New York Herald,* where the story appeared under a glaring headline on the front page, above the fold, with pictures of Hornaday and the Audubon Society's Mabel Osgood Wright. "I think it is eminently justifiable and right that the Public should know the facts in this unfortunate case, and for my part I am taking no pains to conceal them," Hornaday wrote with understatement to his friend John Phillips. Hornaday's offensive caught the Audubon Board of Directors completely off guard. His aggressive actions through public interviews and private letters shaped the public perception of the story and threw the Audubon leadership on the defensive from the very outset. They never recovered their composure. Hornaday publicly revealed for the first time during an interview that he had been approached in 1906 by gun makers with offers of cash "to purchase peace with the Zoological Society." Despite Winchester's claims to benevolent motives, Hornaday established a narrative that the gun makers wanted to use their blood money to negatively affect wildlife conservation. George Bird Grinnell, who voted to accept the money at the Audubon Board meeting, considered the whole affair a tempest in a teapot that would soon blow over. "The criticism received has been confined to the New York papers," he wrote to board member Joel Allen, "and has had very little currency outside of the city."[4]

Pearson buckled under the pressure despite Grinnell's efforts to steel his resolve and lobbied to reverse the board's decision. "In my opinion we have almost everything to gain and very little to lose by rescinding our former action," the secretary wrote Joel Allen on June 14. "The gift was accepted by the Board at a meeting on June 2nd on a condition which it has been found impossible to comply with," he wrote to the board's Frank Chapman, who was out of the country at the time of the vote. Chapman was to play a key role in the Audubon Society's reversal of its decision. His influence was substantial, and his absence at the vote provided an excuse to call for another one. The ploy worked as well as Pearson had hoped. Two weeks after voting to accept the gun makers' money, the Audubon Society's Board of Directors rejected it. In a *Bird-Lore* editorial explaining the decision to the membership, Pearson bluntly stated that the National Association of Audubon Societies (NAAS) did not exist to protect birds today only to shoot them tomorrow.[5]

"Of course, Hornaday and his friends have won a triumph," George Bird Grinnell wrote in disgust to William Brewster, cofounder of the American Ornithological Union. He took the liberty of advising Madison Grant to restrain Hornaday's tendency to gloat. But it was a pyrrhic victory. Although the Audubon Society reversed its position, it never adopted Hornaday's agenda, nor did the controversy revive his moribund campaign to restrict the use of automatic and pump shotguns in New York State, as he had hoped. In fact, he would battle Audubon again on that issue and others throughout the 1920s and 1930s. His quickness to resort to planting stories in the *New York Herald* and other yellow journals and spinning the whole Audubon controversy to his advantage ruptured his relations with many conservationists. Grinnell, for one, appears to have lost much of his respect for Hornaday during the incident. Their relationship continued to deteriorate afterward, to the point of open and public hostility.[6]

Pearson was a little more forgiving and worked with Hornaday in the immediate aftermath of the Audubon Society's reversal, before their relationship also disintegrated during the tumultuous 1920s. Pearson, however, misunderstood Hornaday's motives. In his memoirs, *Adventures in Bird Protection,* a less-than-repentant Pearson wrote that the crisis was caused mainly by men not connected with the NAAS and that "some of it was from a man who, we had reason to believe, was chagrined because the money had not been given to him to spend." This statement, clearly referring to Hornaday, missed the point of Hornaday's principled stand. Hornaday really believed that conservation and gun makers' blood money did not mix, as the next twenty-five years would demonstrate. And Pearson failed to understand Hornaday's utter frustration at Audubon's lack of support for his shotgun reforms. Hornaday had made it clear to Leonard and the Winchester company in March that he would not accept their money unless they agreed to his precondition on the number of shells their shotgun magazines could hold. He did not regret their refusal. He already had his discretionary fund for wildlife-protection efforts in place, with plans to vastly expand it. With friends and benefactors like Andrew Carnegie, George Eastman, Henry Clay Frick, and Olivia Sage, among others, he simply had no need for the gun makers' money with its nasty strings attached.[7]

In another sense, though, Pearson hit closer to the mark. Hornaday's position, however principled, was indeed a personal battle, and he waged it vindictively. He considered the Audubon board's vote to accept gun money a slap in the face and could not believe that the directors would accept money from the shotgun manufacturers at the very time he was trying to restrict the use of their product. In one of his characteristic egocentric moments, he imagined himself as central to the board members' considerations when he analyzed their votes. "General hatred of W. T. Hornaday" was his explanation for how Joel Allen voted. He then ascribed the ballots of two other board members, including his friend and best man at his wedding, Frederic Lucas, to their being Allen's lackeys. He believed Grinnell's support of the measure had to do with his long friendship with the owner of the Remington Arms Company. To Pearson, he attributed the least charitable reason: "purchased outright with $3,000 per year of gun-maker's money." He saw only Mabel Osgood Wright as acting for a positive end—to secure more funding for *Bird-Lore*.[8]

The Audubon battle was his first bruising battle with other national conservationists. Possibly he could have achieved his objectives with more tact and less publicity, but Hornaday sought to shame them publicly. In his mind, it was preposterous that the Audubon board even debated the gun subscription. It was as if the Women's Christian Temperance Union had voted to accept money from a whiskey distiller on the condition that they limit their campaign against alcohol, while continuing their educational reform. He did not see himself trying to correct an error, but rather as being in a life-and-death battle for the very soul of wildlife protection. With such important stakes, he could not afford to give quarter.

The gun makers, led by the marketing department of Winchester, did not give up on their plan to influence wildlife conservation from within. After the Audubon fiasco, they created their own organization, the American Game Protection and Propagation Association, although they soon dropped propagation from their name (AGPA). John Burnham of the New York Forest, Fish, and Game Commission (NYFFGC) was hired to lead it. At first, Burnham did not want the job. "In fact when it was first broached to me," he wrote to New York game commissioner James Whipple, "I turned down the proposition." But he eventually relented. A blue blood in the mold of Grant and Grinnell, Burnham possessed solid credentials to lead a sportsmen's organization. In addition to his service on the NYFFGC, he had spent seven years as the business manager of *Forest & Stream* magazine under George Bird Grinnell, where he received a first-rate education in the sportsman's ideal for wildlife conservation. Burnham also traveled and hunted extensively in the United States and Canada and was thus familiar with game conditions throughout the continent.[9]

Hornaday and Burnham already had a rocky relationship before Burnham accepted the gun makers' blood money. In 1910, amid a fight on a bill to protect gray squirrels in New York State, Hornaday's disparaging remarks about Burnham stung the recipient so badly that he vowed never to forget them. "Such action was inhumanly cold-blooded," Burnham wrote, "and I lost my respect for the man." A year later, in a letter to him, Hornaday sternly criticized Burnham's management of game enforcement on Long Island. Although Hornaday calmly asked that his recipient not take the letter personally, he did call Burnham's administration of game laws "a lamentable failure" and asked, "how can you expect any one in possession of the facts to believe that you are a strong and capable official?" That was vintage Hornaday. Despite their past differences, though, Hornaday's main issue in the fall of 1911 was that Burnham accepted the leadership position of the gun makers' organization. To Hornaday, the AGPA was a "commercial" organization designed to protect the products of its founders, the pump and automatic shotguns that he was trying to limit. Even worse, he feared, it might actually imperil nongame species by diverting all conservation energies toward measures that protected only the species valued by sportsmen. From now on, he had to be a vigilant sentinel on guard to protect the entire conservation movement from being steered in the wrong direction by the gun makers and their henchmen at the AGPA.[10]

Hornaday's first opportunity to challenge the AGPA occurred at the December 1911 convention of the New York Fish, Forest, and Game League in Schenectady. He personally led the campaign against the AGPA candidate for president of the league. His hostile attitude incensed Burnham, who denounced Hornaday's conduct in the *New York Times*. "Dr. Hornaday is altogether too prone to criticize unfairly everyone who does not agree with him," Burnham wrote. Grinnell realized Burnham had his blood up, but cautioned him from responding to Hornaday publicly. "It is partly for this reason that I thought it bad judgment for you to reply to Hornaday," he wrote. "The making of such a reply gratified your personal feelings but did no good to the cause of game protection, and will likely stir up another reply from Hornaday." Burnham ignored Grinnell's advice and responded with hyperbole that exceeded Hornaday's own. Burnham's candidate,

George Lawyer, won the election. Hornaday took comfort in the belief that he had attracted so much negative publicity upon the AGPA that it would prevent Burnham's minion, Lawyer, from fully following the line of the gun makers. Hornaday's effort was only the opening shot in what would become a twenty-year war.[11]

After the meeting, Hornaday and Josephine took a greatly needed vacation to Bermuda, where they enjoyed the glass-bottom boats and warm breezes. No doubt, Hornaday recalled the long-gone days of his youth hunting across the tropics in two continents.[12]

He returned from his restful vacation full of fire to revive his stalled campaign to limit automatic and pump shotguns. "I think we can expect some results from the seed sown last year," he wrote to Henry Fairfield Osborn in January 1912. George Bird Grinnell attempted to persuade Madison Grant that Hornaday was bringing discredit to the society. In February 1912, Grinnell wrote Grant to tell him that "Hornaday's judgment is wrong" on the subject of shotguns. "I imagine that you and Osborn have, unconsciously, been influenced by Hornaday's rantings about his favorite bugaboo," Grinnell wrote. "In my opinion he is an unsafe man to tie into these matters." Even if he wrote nothing on society stationery, Grinnell reasoned, Hornaday embarrassed his employers simply by virtue of his position as director of the New York Zoological Park. But Grinnell's attempt to influence Grant and Osborn to restrain the director did much less to doom Hornaday's most recent shotgun crusade than did the fact that the issue simply had no traction in the state legislatures. By October, he decided to cut his losses. "My fund is much too low to admit of the expenditures of anything more on the automatic-gun campaign," he wrote with obvious disappointment to George Shields. It would have to wait another day.[13]

Perhaps his shotgun campaign never gained sufficient steam for the simple reason that many other, more practicable matters distracted him. Early in the year, he was back in the thick of the seal fight, taking on all the members of the Fur Seal Advisory Board, a panel of experts advising the U.S. Commerce and Labor Department including Charles Townsend of the New York Zoological Society's Aquarium. In July 1911, Great Britain, Japan, Russia, and the United States had signed a treaty ending pelagic sealing. It was a major achievement in diplomatic conservation that ended a destructive practice likely to result in the extermination of the species. The end of pelagic sealing saved the seal from following the path of the buffalo, if not the passenger pigeon and great auk, in the twentieth century. Yet, Hornaday was not content with this victory, and he pressed for a closed season on land. A clause in the agreement reserved a right for America to enact a closed season on land killing under the proviso it compensated Canada and Japan. As Henry Wood Elliott lobbied for a closed season in Congress through the sensationalistic hearings before the Rothermel Committee, Hornaday took on the Commerce and Labor Department's advisors who claimed such a rash action would actually do great harm to the seal herd.[14]

The scientific experts of the Fur Seal Advisory Board, led by a duo from Stanford University, David Starr Jordan and George A. Clark, flatly opposed any closed season on land for two critical reasons. On diplomatic grounds, Clark, who was the spokesman for the experts, argued that any closed season violated the spirit of the treaty ending pelagic sealing and would lead to "the resumption of pelagic sealing and the ultimate destruction of the herd." Zoologically, the experts asserted that it would be dangerous not to harvest the surplus bachelor males. Older males dominated the herd with as many as thirty females in their harems, but there were epic battles for control of these harems. To the seal experts, an excess bachelor population only meant more fighting and the likelihood that warring males would trample the defenseless pups. "Fighting may be seen anywhere in the rookeries and many of the very young seals are trampled to death," Charles Townsend wrote, fearful of more aggressive seals on the rookery. Hornaday was frustrated by the society's stance on this issue, especially after Osborn sided with aquarium director Townsend in denouncing the closed season on land. It, Osborn wrote to Congressman William Sulzer of New York, "would exterminate the great seal herd of the United States." Dismayed that Osborn sided with the opposition, Hornaday informed Elliott "that fool theory has actually enlisted the adhesion of no less a man than Henry Fairfield Osborn—who claims to believe sincerely that the surplus seals need to be killed for the good of the herd."[15]

The trampling theory positively enraged Hornaday. He considered it a demonstration of human arrogance and bad science. "There is not a case on record where man has successfully stepped into regulate the breeding of a wild species with successful results," he told a congressional committee. He could not believe that the men of the Fur Seal Advisory Board could be as ignorant as that. Instead, he convinced himself that they were too vested in the commercialization of the seals to see the stupidity of their actions. "I am out of all patience with those namby-pamby 'experts,' who never yet, up to this date, have accomplished one good purpose for the cause of the Fur-Seal," he wrote to a sympathetic Elliott, "and who have become frantically obsessed with the desire for killing. Hereafter they will belong in the ranks of the pelagic sealers and the killing gang." Hornaday would later draw similar connections between the U.S. Biological Survey and the hunting industry. Although he stood almost alone among the scientific community in denouncing the trampling rationale, his voice proved more persuasive with politicians. In August 1912, Congress enacted a five-year closed season.[16]

Hornaday's victory both dumbfounded and greatly angered the seal experts. His confrontational style irked them, but they found his direct challenge to the Progressive ideal of reliance on scientific experts even more troubling. Jordan wrote in his memoir, *The Days of a Man*, that congressional approval of a closed season was yet one more example of the "failure of a certain class of officials to take advantage of expert knowledge." Despite the dire predictions emanating from Stanford University, however, the closed season did not spell doom for the seal. When the closed season expired in 1917, the seal population had grown to almost a half million animals. By 1920, seal pelts harvested on the Pribiloff Islands contributed more than $1 million annually to the federal treasury. Later research showed that the deaths that Clark, Jordan, Townsend, and others attributed to trampling resulted, in fact, from a hookworm. Yet Hornaday claimed too much credit for the role played by the closed season in the fur seal's sudden population growth. The historic ban on pelagic sealing was also a determining factor in the increase.[17]

There was still one more act to be played out in the drama of the fur seal. George Clark pressed for a repeal of the closed season. "I must request you to cease your utterly useless agitation for a repeal of the close season law," Hornaday wrote Clark in February 1913, "accept your defeat gracefully, and give the new law a fair deal." Clark, however, had no intention of relenting and made the mistake of dashing off a threatening letter to his nemesis in the Bronx. And, like so many before him, he had cause to regret it. "The recent 'mess' in the fur seal situation was too strongly and too definitively influenced by your reputation as a champion of wild animal conservation to permit of your escaping responsibility too easily," Clark wrote to Hornaday, adding, "I will make you smart for having meddled in the fur seal matter." A Stanford University fur-seal expert was a small fish in Hornaday's pond, but he still retaliated, holding the correspondence for almost a year before cleverly releasing it to Representative John Rothermel just before Clark was scheduled to appear before his committee. Much to his embarrassment, Clark spent a good deal of his testimony on the defensive as Rothermel relentlessly quizzed him about the exchange of letters and the intent behind his threat to a fellow scientist.[18]

As he was enjoying the fruits of his successful effort to obtain a land-based closed season for the fur seals of the Pribiloff Islands, Hornaday was entering the ongoing campaign to pass a law protecting migratory birds. Earlier in the year, he had declined to participate in the lobbying effort, but his interest revived in September when Representative Frank Mondell of Wyoming, a notorious opponent of all conservation, succeeded in striking the bill from the calendar. It "is not likely to become a law within the time of living men," Hornaday wrote pessimistically.[19]

Hornaday faulted the bill's chief lobbyist, John Burnham, for this catastrophe. He believed the fatal flaw was Burnham's determination, and by extension the gun makers' determination, to protect only game species of migratory birds. Hornaday considered it class legislation favoring only one group of people, hunters, and, worse, a thinly veiled effort to subsidize the producers of automatic and pump shotguns. He wanted instead to protect all migratory birds, including songbirds generally not hunted as food. Not only would this dramatic shift in focus increase the scope of the protection, it would also expand the number of people who benefited from the passage of the bill. To the end of salvaging the situation and recasting the bill, Hornaday called together a dinner of conservation notables representing the most important and influential wildlife-protection organizations. "The chief promoters of the hopeless bill grasped the idea, as a drowning man does at a straw, and all save one of the persons invited graciously accepted," Hornaday recounted in *Thirty Years War for Wild Life*. That one person was John Burnham, who actually attended part of the meeting after initially declining because of a schedule conflict.[20]

At the dinner, held at New York's posh Century Club on September 18, 1912, the participants agreed to amend the bill to protect all migratory birds, as Hornaday urged. Not everyone was satisfied with the outcome. Burnham felt cheated and steamrolled by the result of the vote, but powerless to buck it. He enjoyed the consolation of remaining the bill's chief lobbyist in Washington, though, and the consortium of conservationists rejected Hornaday's effort the following month to replace Burnham in that post.[21]

On October 1, Hornaday delivered an illustrated lecture advertising the merits of the rewritten legislation before the Fourth National Conservation Congress in Indianapolis, Indiana. As chairman of the congress's Committee on Wildlife Protection, he ensured that the congress endorsed the revised Weeks-McLean bill, as the eventual legislation protecting migratory birds became known. "The only way in which all these valuable migratory birds can be saved to us is through the strong arm of the National Government," Hornaday stated, "and a Federal law for the protection of all migratory birds!" He also rolled out a justification for the bill that he hoped would appeal to key constituencies of American society. "The destruction of our insect-eating birds means a great increase in the armies of destructive insects, a great decrease in our agricultural products, and a great loss to consumers and to farmers." Using research the Biological Survey had conducted for over a decade, he stressed that protecting insectivorous migratory birds would save the U.S. economy tens of millions of dollars by reducing enormous losses to agriculture, which would benefit both consumers and farmers. Unlike that of the Fur Seal Advisory Board, Hornaday found the evidence and conclusions of the Biological Survey overwhelmingly convincing. Others, like William Dutcher of the National Association of Audubon Societies, had made this argument to the public in recent years, but none had made it the centerpiece of a national wildlife-protection campaign or with such vigor before.[22]

When he was not working on seals, shotguns, or migratory birds, Hornaday was writing a book to garner public sympathy for the cause of wildlife protection. "I have been so profoundly impressed by the destruction of wildlife on the continent of North America, and the enormous resultant loss to the industries and economics of this country, that I have felt compelled to write a book for the purpose of arousing a great mass of the people who now take no interest in the subject," Hornaday wrote to Henry Fairfield Osborn. Such efforts after putting in full days at the New York Zoological Park in the Bronx greatly depleted Hornaday's vast stores of personal energy. "As a matter of fact," he wrote to fellow conservationist Henry Shoemaker of Pennsylvania in September 1912, "I am tired enough all the time but after I have written at my game protection book as long as I can keep awake, I am too tired to do anything but to go to sleep."[23]

Our Vanishing Wild Life, the result of his labors, was William Temple Hornaday at his best and most passionate. His language was strident, moralistic, and full of both condemnation and humor. "Its burning and indignant pages remind me," wrote a Sierra Club reviewer, "of the zeal of the old anti-slavery days when the force of great moral convictions won the day against greed and wrong." If he failed to inspire American citizens to join what he referred to as the "Army of Defense," then he would shame them into it. Full of graphic photographs and charts, Hornaday brought the images and costs of the slaughter into his readers' living rooms. Philosophically, *Our Vanishing Wild Life* followed the well-worn path Hornaday had already trod in *The Extermination of the American Bison* and *The Destruction of Our Birds and Mammals:* Namely, that humans killed wildlife at a rate far in excess of the animals' ability to reproduce; that the commercialization of animal products, such as feathers and meats, vastly fueled the slaughter of wildlife; and that ever-improving technology allowed humans to kill animals at ever-increasing rates. The future for wild animals, Hornaday lamented, looked dim indeed. He stressed that regulations and laws were only one part of the solution. The other, perhaps greater need was for the American people to shake off their apathy and take an active interest in their great natural heritage. Hornaday argued that all Americans bore a moral responsibility to protect the nation's wildlife because it belonged to no single interest group. Echoing the theme of immediacy that Hornaday would have heard as a young boy at Adventist revivals on his father's farm, he demanded instant atonement and immediate corrective action.[24]

A subscription fund managed by the New York Zoological Society paid the cost of publication and distribution of *Our Vanishing Wild Life*. Hornaday's favorite benefactors, Andrew Carnegie and Olivia Sage, donated one thousand dollars each to a ten-thousand-dollar fund that allowed the society to mail more than seven thousand copies gratis to policy makers involved in wildlife-protection policy. "The book has now gone to *every* member of *every* legislature now sitting in the United States, and every member of Congress," Hornaday wrote with some satisfaction to Theodore Roosevelt in early February 1913.

"It will also go to every governor, Supreme Court Judge, game commissioner and state game warden." When Hornaday testified before the House Ways and Means Committee at the end of January 1913 on behalf of a clause in a tariff revision banning the importation of plumage, he referred congressmen to their recently delivered copies of *Our Vanishing Wild Life*.[25]

Reviewers praised the book. "The true sportsman, the nature-lover, the humanitarian—in short, all good citizens of all types—should read this book and should respond to the appeal Mr. Hornaday makes," Roosevelt, hardly an impartial reviewer, wrote in *Outlook* magazine. Yet Hornaday ran some risk with those who thought like the former president by including "the Gentleman Sportsmen" in "the Regular Army of Destruction." After spending six paragraphs describing the vast contributions sportsmen made to wildlife conservation, he concluded, "For all that, however, every man who still shoots game is a soldier in the Army of Destruction!" Another sportsman, Aldo Leopold, a young forester in New Mexico who would become one of the twentieth century's towering environmental figures, also looked past this comment. In his review of Hornaday's book in *Pine Cone* magazine, Leopold called it "the most convincing argument for better game protection ever written."[26]

Some were quick to crown *Our Vanishing Wild Life* a seminal work of conservation writing. "It is by all odds the most comprehensive and convincing presentation and discussion of the subject that has ever been produced," wrote George Gladden of *American Review of Reviews*. This was no doubt true, as it was one of the first books devoted exclusively to wildlife protection. It had a significant impact on the evolution of Aldo Leopold. Then recovering from nephritis at his parents' home in Iowa, he read Hornaday's book in his sickbed. The strident style, demand for individual action, and emphasis on communal responsibility had a profound impact on Leopold. According to his biographer Curt Meine, *Our Vanishing Wild Life* was "the book that had the greatest effect on Leopold." It shifted his thinking from game animals toward all wildlife, and instilled in him a sense that humans bore moral responsibility for the animal kingdom.[27]

As the printers churned out copies of *Our Vanishing Wild Life*, John Burnham maneuvered to get the Weeks-McLean bill protecting migratory birds up for a vote. "This is a very large step in wildlife protection," George Bird Grinnell wrote to Burnham in January

1913, "perhaps the largest that has ever been made in this country." In February, however, the Weeks-McLean bill got stuck in a parliamentary logjam. The Rules Committee would not report on it because if it did, it would bring up the next item on the calendar, a contentious workman's compensation bill. In less than a month, a new Congress would be seated and a new president inaugurated. After a failed effort by Congressman Oscar Underwood of Alabama to call out the bill, Burnham resorted to a classic conservation ploy, one used to create the forest reserve system in 1891 and the Wichita bison preserve in 1907. He persuaded a distant cousin, Senator Henry Burnham of New Hampshire, lame-duck chairman of the Senate Committee of Agriculture and Forestry, to add the provisions of Weeks-McLean as a rider to the annual agricultural appropriations bill. In the closing hours of his administration, President William Howard Taft warned Congress not to try any funny business with the appropriations bills and threatened to veto any suspicious provisions. Taft was already on record as questioning the constitutionality of the federalization of bird conservation. In the rush to sign all pending items on his desk, however, he failed to read the agricultural appropriations bill before signing it. He later commented that he would not have signed the bill had he known of Burnham's handiwork.[28]

Despite the apparent comity within the wildlife-protection movement over the need to protect migratory birds and the collective realization that cooperation resulted in the successful completion of what Pearson called the "most gigantic single campaign ever waged for a bird protective bill," bickering over what to do next started almost immediately. Burnham recommended the formation of an advisory panel, an idea that played well to the prevalent Progressive Era ideal of management by a commission of recognized experts, as the historian Samuel Hays depicted so well in his work *Conservation and the Gospel of Efficiency*. More valuable to Burnham, it would give sportsmen a strong voice in the regulations, thus undermining Hornaday's move to minimize the role of hunting in shaping the legislation. Burnham was concerned that the effectiveness of the legislation rested on some measure of cooperation from hunters in the field. Thus he wanted input from those affected by the laws and requests for restrictions to come from those who would be most affected by them. "It will be much better to have requests for the curtailment of privileges come from persons affected by such curtailments other than to put it over them by force and arms," Burnham wrote his mentor George Bird Grinnell.[29]

Theodore Palmer and A. K. Fisher of the Biological Survey drew up a blueprint for protective regulations and allotted three months for public comment. Hornaday was taken aback at how much Palmer's regulations reflected the desires of the sportsmen. He had always considered Palmer one of the good guys in wildlife conservation. "Dr. Palmer is a man of incalculable value to the cause of protection," he had written in *Our Vanishing Wild Life*. A mere three months later, he described Palmer as an "enemy." "And after I have been pinning medals on Palmer for the past five years, and came to his rescue—at his request—when the Biological Survey was attacked in the Appropriations Committee two years ago," he wrote in disgust to George Shields.[30]

Secretary of Agriculture David Houston adopted the idea of a board to advise him on migratory bird protection regulations, and he named John Burnham chairman. In June, Palmer prepared a preliminary list of members and named George Bird Grinnell as the representative of the New York Zoological Society. Hornaday was aghast. "People may do what they please about recognizing me individually, and it makes no difference to me," Hornaday wrote to Henry Fairfield Osborn, "but for any man to appoint George Bird Grinnell to represent our Society is a distinctly unfriendly act, and should be treated as such." Hornaday closed his short letter with the statement, "I would not accept the position on the Committee,—even if made."[31]

Six weeks later, on July 21, Secretary Houston formally announced the creation of his advisory committee and named fifteen members, including William T. Hornaday of the New York Zoological Society. Hornaday reneged on his private declaration to Osborn that he would not accept a position on the committee, but he was hardly an active member. He attended only one of the board's annual December meetings between 1913 and 1925.[32]

The Weeks-McLean bill absorbed only part of Hornaday's attention in 1913. He devoted more energy to lobbying for the inclusion of the ban on the importation of plumage in the tariff revision. Millions of egrets and heron died needlessly so that their plumes could adorn the fancy hats of American women. In 1886, Frank Chapman had counted over 700 such hats, 542 of which were decorated with bird plumes or contained stuffed birds, during two afternoon strolls in

New York City. Over the next two decades, several states had adopted bans on killing certain birds or selling plumage, but they were poorly enforced. In 1903, a salaried Audubon warden, Guy Bradley, was murdered by market hunters for his attempt to enforce such a ban enacted in Florida two years before. As the method of last resort, conservationists had resorted to shaming women for their blood thirst. At the time that Chapman made his fancy-hat survey, John Burroughs condemned the "barbarous taste, which prompts our women and children to appear on the street with their head gear adorned with the scalps of our own songsters." Charles Dudley Warner, who cowrote *The Gilded Age* with Mark Twain, was even more pointed: "A dead bird does not help the appearance of an ugly woman, and a pretty woman needs no such adornment." Hornaday himself had written rather caustically in 1897, "One of the strangest anomalies of modern civilization is the spectacle of modern woman—the refined and tender-hearted, the merciful and compassionate—suddenly transformed into a creature heedlessly destructive of bird life, and in practice as bloodthirsty as the most sanguinary birds of prey."[33]

Henry Oldys, a Maryland doctor and Audubon Society member, recognized that the Democratic Party's victory in the 1912 election might prove beneficial for wildlife. Since its inception, the party of Jefferson had opposed protective tariffs, and President-elect Woodrow Wilson promised a thoroughgoing reform of the duties. Oldys believed it would be possible to prevent the importation of plumes. Although many of the egrets and herons were killed in Florida or Louisiana, they were shipped to Paris, where they were turned into fashionable items and then imported into the United States as millinery. Oldys offered his ingenious plan to Audubon's T. Gilbert Pearson, who promptly contacted Oscar Underwood, the Alabama Democrat who had the daunting task of shepherding the massive tariff reform measure through the House. Underwood summoned Pearson to appear before his committee at the end of January, and Pearson wisely invited Hornaday to testify as well.[34]

According to Pearson's account, the two men discussed strategy that evening and agreed to ask for a total ban on the importation of all bird species' plumes based on New York's Shea law. "I think the boldest position is the best one!" Hornaday exclaimed to former president Roosevelt. A ban on only American species would serve as the fallback position, Pearson recalled. Hornaday's testimony, on January 30, 1913, resembled his rhetoric in *Our Vanishing Wild Life*. "The demands of the feather industry are absolutely inexorable and remorseless," he told the committee. He also heaped some scorn on the consumers of the plumes, criticizing the "Smart Set" for fueling the slaughter in the first place. When one committee member asked, "Where would the ladies get their hats? They seem to be necessary, and every woman has a feather on her hat, even if it is a rooster feather," Hornaday responded that a thriving ostrich industry could supply all the plumes necessary without any threat of extinction, if only the "Smart Set" condescended to wear them.[35]

Pearson took a slightly different tack, addressing the national implications of the issue and appealing to the Progressive Era sense of rationalism by stressing the large number of harmful insects these birds consumed and the resulting benefit to the nation's food supply. He also called special attention to the fact that several states already prevented the sale of plumage and the killing of certain birds that provided the plumes so desperately sought by the milliners. When one committee member asked if it was legal to kill these birds in the southern states, Pearson replied in his North Carolina drawl, "It is illegal to kill them or to sell their feathers in virtually all the States." "But the burden of proof is on us to show that these birds were killed in the State," he added. A national law would protect the rights of these states, Pearson argued, in the hope of winning over fellow southerners to an increase in federal power.[36]

Both Hornaday and Pearson were optimistic after the hearing. Hornaday described the mood of the committee to his friend Theodore Roosevelt as "not only keenly interested, but also sympathetic." "When after the hearing we took our hats and departed," Pearson wrote, "we felt pretty sure that our proposition would have favorable consideration." Francis Harrison, a New York Democrat, advised Hornaday that a ban on the importation of only American bird species' plumes seemed more likely to pass the committee, but welcomed Hornaday to write a stronger provision for their deliberation. Despite Harrison's misgivings, the stronger version carried the committee, and the House passed it on April 7, 1913. "Before the milliners were awake to what was being done," Hornaday wrote to one correspondent, "our clause was actually in the tariff Bill, safe and sound. Now they are begging for mercy."[37]

In late April, Hornaday and Pearson met several times with representatives of the millinery lobby, a fact both decided to omit from their memoirs. Hornaday recommended to Madison Grant that the New York Zoological Society work with the businessmen and add a clause to the legislation allowing for the importation of game birds' plumes. "I think that the concession would not lead to *any* slaughter of *American* game birds, for millinery purposes,—chiefly for the reason that all save a very few of our states prohibit the sale of such birds," Hornaday reasoned. Grant accepted the concession, "so long as their use by the feather trade does not constitute a special menace to those birds, and put an extra premium on their destruction." This attempt to broker a compromise failed, however, as the millinery representatives took a gauge of the Senate and found it friendly to their industry.[38]

Senator Hoke Smith, an influential Democrat from Georgia, acted on behalf of the milliners in the upper chamber. In May, the Senate Finance Sub-Committee called for hearings on the tariff bill in the middle of the month and gave the conservationists only two days' notice. Pearson was unable to attend, and Hornaday raced to Washington with little preparation. He gave rambling testimony on the general need for bird protection. Hornaday was always a nervous public speaker, but it never deterred him from making the effort. At one point, Senator Smith asked Hornaday the value of the silver pheasant of Malaysia that he was using to illustrate a point. When Hornaday replied that it was $3.50, Smith asked for its value to society, not its commercial value. "The value of all birds of the world that are not to be killed and eaten by man is sentimental," Hornaday responded. Smith replied that he thought the value was in the bird's ability to consume noxious insects. "Most decidedly," Hornaday responded. "I had not come to that. That is a very important fact." After confusing rounds of discussion about crows and quail, Hornaday essentially dismissed himself, reminding the committee that others were waiting to testify. Over the course of his career, he made several effective performances before legislative committees, but this was not one of them. Clearly, he was nervous and bombed. Resting on sentiment to justify the protection of plumage birds, he strangely neglected the much stronger position he was concurrently using to promote the Weeks-McLean Act, that of birds as nature's defenders against harmful insects.[39]

Hornaday did go on record as being firmly opposed to an amendment advocated by Senator Moses Clapp, a Minnesota Republican. Clapp wanted to amend the House version of the legislation to allow the importation of any plumes of birds considered edible or pestiferous. "The proposals to admit the plumage of game birds and of birds killed as pests are both utterly inadmissible," Hornaday, who no longer advocated a concession on game birds, wrote in an editorial. "Think of it!" Pearson wrote exasperatedly. "The feathers of any bird which any one eats at any place on this green earth, or any bird which any interested person may be pleased to catalogue as a pest, may be imported under this provision!" But the Senate passed the Clapp amendment, changing the entire outlook of the ban. "Altogether, this is the ugliest situation that I ever have come up against in thirty years of legislative experience in behalf of wildlife," Hornaday exclaimed to a cousin in June. He issued a public statement on June 20 describing how the millinery industry had "captured" the senators who favored the Clapp amendment. His use of this derogatory term earned him a rare rebuke from the *New York Times* editors. "The Senators have not been 'captured,'" the editors wrote, "they have merely allowed themselves to be over persuaded by plausible arguments, most of which, though false in part or whole, are usually made honestly enough." The *Times* urged Hornaday to get back to confronting the arguments of opponents of the ban and to tone down the ad hominem remarks. In July, he appraised the desperate situation to Henry Fairfield Osborn using one of his patented Civil War metaphors. "In the United States Senate, our cause is now at 'the bloody angle' of Gettysburg," he opined. "Thus far the enemy are in possession of the field; but in the end, 'we will lick them out of their boots.'" This, as it turned out, proved a very accurate analogy.[40]

Hornaday and Pearson unleashed a blitzkrieg with vigorous editorials in *Bird-Lore*, fiery public letters, and appeals to all those interested in conservation to remain vigilant. In a letter to the American people in early August, Hornaday portrayed the situation this way: "Shall the two or three dozen New York importers of wild birds' plumage be permitted to defeat the will of two or three dozen millions of American people who abhor the traffic and desire its discontinuance?" Hornaday publicly named three senators he considered to be the main lobbyists for the industry within the upper chamber. Despite the appeal of the *Times* to tone down his peppery rhetoric, Hornaday considered such tactics necessary to achieve a plumage ban. "Victories are won by sledgehammer methods," Hornaday wrote to Henry Oldys in late July. In one of their clever uses of hardball tactics, Hornaday and Pearson showed a film prepared by Edward McIlhenny, a wealthy Louisiana sportsman who earned his money producing Tabasco sauce, showing the killing and skinning of egrets and herons to congressmen and others in Washington. At the conclusion of the film, Hornaday and Pearson addressed the stunned audience. Attempts to arrange a private meeting with the president were unsuccessful.[41]

The most effective lobbyist on behalf of the plumage ban, and perhaps the one most responsible for obtaining the president's support of the House version of the bill, was virtually unknown on a political level among the wildlife conservationists—the First Lady of the United States, Edith Axton Wilson. While on vacation in New Hampshire with their daughters, Mrs. Wilson sent her husband a list of suggestions on the tariff revision, including full support for the House version of the plumage ban. "I shall of course do what I can to get the duty on art and the objectionable provision about the birds taken out of the bill when it gets into the conference," the president wrote to his wife. "I have not yet had a chance to take the matter up; and this is not yet the moment." The president would prove good to his word.[42]

On August 16, Senator George McLean, the Connecticut Republican who had cosponsored the great migratory bird bill with Representative John Weeks, introduced an amendment to the Senate bill so that its language on the plumage clause matched the House version. The Senate concurred in early September, and President Wilson signed the Underwood-Simmons Tariff Act on October 3, 1913. Wilson achieved a remarkable political success, and the conservationists enjoyed theirs as well. "It is a great victory," George Bird Grinnell wrote to Pearson, "and one on which we may all congratulate ourselves."[43]

Hornaday, however, was less willing to claim this as the victory of a movement. Instead, he understood it as the herculean accomplishment of only two men representing a pair of organizations: himself of the New York Zoological Society, and Pearson of the National Association of Audubon Societies. A little less than ten days after the Senate voted to approve the House language on plumage, Henry Oldys, who had written Pearson in December 1912 to suggest how the tariff legislation could be used to ban plumage, wrote Hornaday a letter. In it he rejoiced in the achievement of a long-sought objective, but suggested that Hornaday took too much liberty in claiming credit for the success. After all, Oldys reasoned, many had testified and other groups had lobbied their representatives and senators. Hornaday took great offense to these remarks. "The above stated is a deliberate falsehood," Hornaday hotly responded, "and unless your mind has ceased to act on rational lines, you knew that when it was penned." To Pearson, he commented that "Oldys is a fool, and on one occasion, he was a liar." Pearson tried to cool Hornaday down. "My feeling is that it is hardly worth while to bother in a public way with the Oldys publicity matter," the Audubon secretary wrote. "He evidently feels that he has been somewhat neglected in the credit matter and possibly under estimates the character and extent of the campaign carried on by our two associations." If Hornaday had once again demonstrated his prickly nature, he at least kept the correspondence out of the public domain, sparing Oldys the unpleasant experiences of Nagel and Clark. He did not mention Oldys at all in *Thirty Years War for Wild Life*.[44]

Hornaday's reward came early the following year, when the French government awarded him a gold medal for his wildlife conservation work, specifically the bird-plumage ban in the Underwood-Simmons Tariff Act. Although deeply flattered by this recognition, he was unable to travel to Paris to be present at the ceremonies. American Ambassador Myron Herrick received the medal on his behalf from a delegation of the French Acclimatization Society. Parisian millinery dealers, angry at Hornaday's role in the tariff ban on their products, cowed French President Raymond Poincare into changing his plan to present the award, or at least that is how Hornaday presented the matter to others. "The feather dealers of Paris hate me royally," Hornaday wrote to his cousin Nina, "and I return their hatred with loathing." That, too, was vintage Hornaday.[45]

The extraordinary year of 1913 had been an exhausting one for Hornaday, who turned fifty-nine in December. "This game protection work is beginning to get on my nerves,—not through the worry at it," he wrote to one correspondent in Jacksonville, Florida, "but through the endless amount of details that I must attend to personally." He had just entered a public brawl with Thomas Carmody, the New York State attorney general, who vowed not to allow state police or courts to enforce the Weeks-McLean Act, claiming that it was unconstitutional. But consolation to his ego came in the form of a ten-page profile entitled "A Champion of Wildlife" published in the *American Review of Reviews*. Although Hornaday had appeared in various biographical directories, this was his first national feature. He was now, without doubt, an influential and powerful national conservation figure with sizable achievements to his credit.[46]

7

THE GREAT WAR

IF 1913 HAD been a whirlwind for William Temple Hornaday, the following year looked to be just as demanding. By the third week of January, the director of the New York Zoological Park already had two months of work piled up on his desk, as he told T. Gilbert Pearson of the Audubon Society. He had to edit the *New York Zoological Society Annual Report,* revise his now ten-year-old *The American Natural History* for Charles Scribner's Sons, and prepare a series of lectures on wildlife conservation for the Forestry School at Yale University, which he later turned into a book. "Well, I am back from the Yale lecture; and I thank the Lord that it is over," he wrote to Josie, who knew how uncomfortable public speaking made her husband.[1]

The Weeks McLean Act had been a tremendous victory for the wildlife protection movement, but the fight was far from over, as Hornaday's public dispute with New York State attorney general Thomas Carmody at the close of the year clearly demonstrated. Hornaday tended to see nefarious economic interests behind the anticonservationist forces. "To them it is business, and they subscribe as business men always do subscribe when their profits are threatened," he wrote in a *New York Times Magazine* piece. Commercial interests found common purpose with a strong, vocal bloc of states' rights advocates and libertarians in the Senate who despised any regulation of personal behaviors. These senators had been asleep at the wheel when the Weeks-McLean language was slipped into the agricultural appropriations legislation the year before, and now they sought to counter the effect of the pernicious new law by starving it of the funds necessary for adequate implementation.[2]

Hornaday had predicted this. As soon as Weeks-McLean had passed, he had recommended a tax on shotgun cartridges as an expedient remedy to provide ample funding and reduce reliance on a fickle Congress for cash. But Madison Grant had nipped the idea in the bud, and it carried no weight with sportsmen. Now Hornaday and other supporters of the act were in a bind. "The present situation is absolutely intolerable!" Hornaday wrote to Senator James O'Gorman, a New York Democrat. *"The people of the country will not stand for it!"* In early May 1914, Hornaday boarded the train to Washington to personally lobby senators for more appropriations for game wardens to enforce the act's prohibition on spring shooting and bag limits. He took advantage of the occasion to present a public lecture at the Shoreham Hotel, one framed by his established progressive narrative of the selfish few damaging the interests of the much larger whole. The millions of "men and boys who are slaughtering our birds are levying a tribute on every American pocketbook," Hornaday told the audience.[3]

In the upper chamber, Senator James Reed, a conservative Democrat from Missouri, led the opposition to funding what he dismissively referred to as "this alleged law," the Weeks-McLean migratory bird act. Like many, Reed considered the law granting the federal government the right to regulate most species of migratory birds a clear violation of states' rights protections in the U.S. Constitution. They expected that the U.S. Supreme Court would vote it down, and stalled for time by withholding the funds necessary for adequate enforcement. On Saturday, May 23, the senator attacked the act as a product of a lobby as powerful and self-motivated as the beef, sugar, and steel trusts. Then he took off the gloves and attacked none other than "that high priest of mercy, Mr. Hornaday," by reading extracts from *Two Years in the Jungle* selected to paint Hornaday as the worst game hog in the world. "I desire to call attention to the fact that, if his heart is now tender and his soul is now shocked at the sight of a dead bird; if he has come to believe that those who occasionally shoot game are monsters engaged in the service of the devil and devoting their lives to acts of atrocity—if that is his present opinion—then, Mr. President, he has undergone a regeneration that has never been equaled since St. Paul saw the light that transformed him from Pagan into a Christian," Reed declared. He added his own colorful commentary after reading each segment of *Two Years in the Jungle* before proceeding to the next one. A chagrined Hornaday mocked Reed in a letter to Theodore Roosevelt afterward: "You should see how he plunged his shining rapier again and again into the devoted bowels of 'Two Years in the Jungle,' and wept and wailed over the dead orang-utans, gibbons, and crocodiles!" Senator McLean wryly noted that Reed's presentation actually generated more sympathy for migratory birds and hence the appropriation. "So I am not wholly bereft," Hornaday commented. Nevertheless, Congress saw fit to allocate only a fraction of the one hundred thousand dollars that Hornaday and other conservationists felt was absolutely necessary to enforce the Weeks-McLean Act.[4]

In November, Yale University Press published Hornaday's lectures at the university's Forestry School under the title *Wild Life Conservation in Theory and Practice*. Although written in a more sober tone than *Our Vanishing Wild Life*, the book echoed the main themes outlined in the earlier book: the duty of the citizen to protect wildlife, the economic value of birds, the costs of extinction, the selfishness of the Army of Destruction, praise of the regulations and laws already attained, and the fundamental right of wildlife to coexist with humanity. Hornaday tailored his argument for the foresters-to-be by calling their attention to the value wildlife protection had for forests. It was not that woods without birds would be silent and dreary, Hornaday said, but that avians protected trees by devouring dangerous insects. Hornaday clearly wanted foresters to pay more attention to wildlife, but he also aimed to influence a larger audience. "Thus far the educators of this country as a class and a mass have not done a hundredth part of their duty toward the wild life of the United States and Alaska," he wrote in the preface to his book. At least Yale had made a step forward; perhaps other schools would follow their example. Charles Christopher Adams, a friend who reviewed the book for *Science*, recognized Hornaday's intent and argued for an even larger readership. The book ought to "be read by every student of zoology and by all interested in general conservation problems," Adams asserted.[5]

In late 1914, with *Wild Life Conservation in Theory and Practice* on the shelves, Hornaday drew up plans for a new wildlife-protection campaign to establish game sanctuaries in the federal forest reserves and national parks. It was an old idea that had been tried before under different guises. In 1902, Congressman John F. Lacey had introduced a game-sanctuary plan for the federal forest reserves to the House of

Representatives, but fiscal conservatives and western grazing interests had blocked the legislation. Although President Theodore Roosevelt had wholeheartedly supported Lacey's bill, his chief forester, Gifford Pinchot, openly opposed it, according to the historian Samuel Hays. Pinchot saw the forests as lands to provide revenue for the government through grazing and logging licenses and fees. Closing part of them to game preserves did not make sense to him. This was precisely the sentiment Hornaday hoped to counter with his lectures at Yale University Forestry School. Instead of bureaucratic competition, Hornaday wanted to unite all the parties around resource protection. The Boone and Crockett Club had revived the idea in 1912, but with such monumental game legislation as the protection of migratory birds on the docket, a sanctuary bill stood little chance of success. When Hornaday introduced his own idea, he made a conscious effort not to repeat the mistakes of the past. He crafted the specifics of his bill to win over the professional foresters, grazers, and other western groups who had opposed earlier efforts. He would take state politics out of the decision-making process and keep it a federal initiative, but avoid conflict with entrenched economic interests by establishing the preserves on recognized wasteland that would not interfere with grazing, logging, or ranching. To win over popular support, Hornaday devised a bold public relations campaign targeting westerners.[6]

On January 12, 1915, Hornaday unveiled his plan at the annual meeting of the New York Zoological Society. With social and economic heavyweights on the Board of Directors, the annual meeting always attracted some measure of media attention, but the attendance of former president Theodore Roosevelt guaranteed even greater press coverage. America's most famous hunter heartily endorsed Hornaday's proposed legislation to place game preserves in the federal forests. "Legislation of this kind is necessary," the reporters quoted him as saying, "in order to give an opportunity for big game shooting to the average man. Wealthy men can have their own game preserves; poor men can't." Roosevelt had remained consistent. Thirteen years earlier, as president, he had favored Congressman Lacey's similar proposal with similar social-justice rhetoric. Hornaday gladly accepted Roosevelt's support, but he never used this particular social justification for the bill himself. Instead of arguing, as Roosevelt did, that the refuges in the plan would provide a safe breeding ground for the propagation of game that would "spill over" into unprotected areas to be hunted,

Hornaday stressed the need of an insurance policy against local extinctions by providing protected, safe havens for breeding. Ultimately, he wanted states to supplement the benefits of his refuge plan by implementing stronger state regulations, such as reduced bag limits and shortened seasons.[7]

The bill progressed slowly over the next several months. On March 4, 1915, Hornaday met in Washington with government officials at the office of Henry Graves, the United States chief forester. Members of the Biological Survey, Grazing Department, and the United States Department of Agriculture solicitor's office were also present. Graves liked Hornaday's idea, but suggested several key amendments to make it more palatable and politically attractive. Hornaday would have preferred to keep the decision in the hands of U.S. Forest Service professionals, not more malleable politicians, but assented to Graves's alteration to place the authority to create the game preserves in the hands of the secretary of agriculture. Hornaday had taken pains to concentrate all the power in the hands of the federal government, but another one of Graves's changes granted the states the right to approve or reject a preserve in their boundaries. Including the states vastly complicated the decision-making process, but Hornaday reasoned that a bad bill would be better than none at all, and he desperately needed the support of the federal agencies to get any legislation through Congress. Graves suggested one amendment that Hornaday considered an improvement. It boldly broadened the bill's scope from Hornaday's preference for areas recognized as wasteland to areas "where they will interfere, to the least extent practicable, with grazing of domestic stock."[8]

Overall, Hornaday was satisfied with the outcome of his meeting with Graves. The revised legislation was more unwieldy in its machinery, but more politically viable, and it was a larger, more expansive bill for protecting wildlife. With a final draft in hand, Hornaday finalized plans for a barnstorming trip of the west to begin at the end of the summer.

Long a student of international affairs and a voracious consumer of news, Hornaday immediately recognized the significance of the world war that was then raging in Europe. In October 1914, he and Josephine took up the cause of Belgian refuge relief after hearing the appeal of Queen Margaret. They raised money and made arrangements to help individual refugees from the tiny war-torn country. These efforts earned a Cross of the Order of the Crown of Belgium from King Albert at the conclusion of the war. Those refugees Hornaday helped and met often repeated stories of German atrocities like those recited in the Bryce Commission report of May 1915, and they deeply impacted his view of a tiny neutral nation literally raped by an aggressive Germany. "To Belgium's calm and pleasant land a Hun gorilla stalked," he wrote in verse. "A club engaged one hairy hand, the other swayed a blazing brand; the earth reeked where he walked." The sinking of the *Lusitania* in May 1915, with the loss of more than 1,100 civilians, including 114 Americans, confirmed his worst fears of Germany's barbaric behavior toward noncombatants, women, and children. In his moral Adventist framework, Germany represented evil, and the Kaiser was no less than Lucifer. Hornaday considered it immoral for America to stand on the sidelines during this epic struggle over the fate of humanity. As was always his view, to be neutral or indifferent in such instances was to ally with wickedness. Doing his own part, Hornaday joined two organizations of the rapidly growing American preparedness movement, the American Defense Society (ADS) and the Army League, of which he was vice president. Within a year, the demands of writing and speaking publicly in support of preparedness and refugee relief had squeezed his conservation activities to a trickle.[9]

On Friday, August 27, 1915, Hornaday arrived in Minneapolis, Minnesota, on the first stop of his speaking tour on behalf of the game-sanctuary bill. From there he continued west, stopping at Denver, Colorado; Cheyenne, Wyoming; Salt Lake City, Utah; Pocatello, Idaho; and Helena, Montana. On September 11, he arrived on the Pacific Coast, speaking in Seattle and Tacoma, Washington, and Portland, Oregon, before crossing into California, where he delivered a lecture in San Francisco on September 16. From San Francisco, Hornaday turned south to Los Angeles and Pasadena. He finished a month on the road in Tucson, Arizona. After a two-week break, he spoke in Albuquerque, New Mexico, on October 13. Local organizations ranging from a sportsmen's association, the Audubon Society, and a state game commission sponsored his lecture. Even the Wyoming Humane Society and State Board of Child and Animal Protection hosted one engagement. The locations varied and included museums, church halls, gymnasiums, YMCA halls, high schools, universities, and the like. Regardless of the room, Hornaday had his stereopticon slides with vivid pictures ready and delivered a dramatic lecture, a primitive version, in a sense, of a compelling PowerPoint presentation. He also read a poem he wrote, "Robbed," which opened with the lines, "Oh, where is the game, daddy, where is the game that you hunted when you were a boy," and closed with, "But Oh! daddy! How *could* you rob me!" Based on the press accounts, Hornaday typically shaped his discussion of the game-sanctuary plan to mirror the objectives of the host organization and addressed local conditions with specific comments on state laws. He demonstrated his always impressive, encyclopedic knowledge on all matters related to the subject of American wildlife. The entire trip went essentially without a hitch; one potentially embarrassing moment was avoided when David Starr Jordan, president of the California Audubon Society who had earlier tangled with Hornaday over sealing, abstained from attending Hornaday's lecture at Blanchard Hall in Los Angeles, even though the California Audubon Society sponsored the event.[10]

Hornaday made one very important acquaintance during his western journey, the young forester Aldo Leopold. In time, Leopold would define the American environmental movement's post–World War II generation with his famous land ethic. On October 13, 1915, however, he was twenty-eight years old and unknown when he sat in the audience raptly listening to Hornaday's lecture at the University of New Mexico. Hornaday had already inspired Leopold with *Our Vanishing Wild Life*, and he lit an even brighter flame when speaking in person. "The sparks that emanated from Hornaday's orations," writes Leopold biographer Carl Meine, "fell on Leopold's own highly combustible convictions." The two men struck up a friendship and would correspond throughout the remainder of Hornaday's life. On the matter of the game-sanctuary bill, Leopold proved an important ally, bridging the gap between sportsmen and Forest Service officers, and helping to construct the cross-interest-group coalition that Hornaday considered necessary for the bill. Leopold formed the Santa Fe Game Protection Association over the winter of 1915–16 with a key objective of building support for the game-sanctuary plan and other game-protection measures. In turn, Hornaday bestowed the Permanent Wildlife Protection gold medal on Leopold "for conspicuous services in awakening New Mexico to wild life protection activities."[11]

Hornaday returned from his western tour in high spirits. "The whole trip was much more successful than I had dared to hope," he wrote to Henry Graves. "The response was glorious," he wrote in *Thirty Years War for Wild Life,* "everything that heart could wish!" Even before Hornaday had returned home, he received an official letter of support from Governor George Hunt of Arizona. "I readily assure you that I am heartily in favor of this movement," the governor wrote, "and that I am prepared to co-operate with you in any way possible." Following this lead, Hornaday sent letters to all of the western governors. By the end of the year, he had endorsements from ten of them.[12]

Governor William Spry of Utah led the opposition. "Vast land withdrawals for various purposes and the extension of Federal supervision within the boundaries of states through congressional enactments has led to a condition which is intolerable to the people of the West," Spry protested. National monuments, national parks, national forests, and Indian reservations locked up over 90 percent of the forest in Utah, according to Spry's calculation, and now William Hornaday wanted more. Of course, Hornaday only wanted to reallocate a small portion of wasteland from land that had already been withdrawn, and he countered by outlining how the sanctuaries would benefit the citizens of Utah. But Spry refused to yield, even when corrected on the exact nature of Hornaday's plan.[13]

Despite the opposition of Governor Spry and the utter indifference in California and Idaho, Hornaday was very optimistic of the bill's chance of success. "I think that no bill ever went before Congress under better auspices, or with brighter prospects, than this one," Hornaday wrote to Henry Wood Elliott on December 30, 1915, "and it would be an awful blow to me if anything should happen to divide the friends of conservation into two camps pulling in opposite directions, and antagonistic over any conservation course." Hornaday was trying to dissuade Elliott from dredging up another accusation of malfeasance by Fur Seal Service that might derail the game-sanctuary plan. After all, that battle had been won. Despite Hornaday's belief that he had enough support for the bill, he decided to employ the parliamentary tactic called the unanimous consent clause in order to bypass potentially messy committee hearings that could suffocate it for years. The drawback of this tactic was that any one congressman or senator could kill the bill on the floor with a single "nay" vote. This made the job of the sponsors, Senator George Chamberlain of Oregon and Representative Carl Hayden of Arizona, a difficult one. They would have to bring up the measure when there were enough votes for cloture but when any avowed opponent was out of the chamber.[14]

To complicate matters further, the Agriculture Department was not done tinkering with the bill. "There are a few modifications in wording which we think will help the bill without, I think, changing its meaning as you had it," Henry Graves communicated to Hornaday through Edwin Nelson. Hornaday failed to prevent these amendments, though he considered them dangerous, as he wrote Senator Hayden, "because any change means complications and delays." In early February, he traveled to Washington to meet with department officials and the bill's sponsors. Although disappointed that government lawyers had altered his pet project, he felt he had no choice but to accept the changes, even if he considered only one of the twenty-one amendments "useful."[15]

Hornaday enjoyed one triumph on his trip to the nation's capital. While in Washington he decided to do some legwork on the Weeks-McLean migratory bird act. Proponents advocated a treaty with Canada or Mexico to both widen the scope of the act and guarantee its constitutionality under Article VI of the U.S. Constitution. Senator Elihu Root of New York, a former U.S. secretary of state, originated this tactic to circumvent a potential hearing on the act before the Supreme Court, and had offered a congressional resolution in January 1913 requesting that the administration enter negotiations on a treaty. Hornaday, of course, favored a treaty, but chafed under its slow progress, which the war had reduced to a crawl. He had written letters and made phone calls, but no one in either the United States government or the British embassy (whose government was acting on behalf of Canada) could provide a coherent account of the whereabouts of a draft of the treaty.

On Saturday, February 12, 1916, Hornaday walked into the U.S. State Department and demanded to know the physical location of the treaty. A clerk pulled the file and found a receipt noting that the treaty had been forwarded to the British embassy in 1915. Hornaday then proceeded to the British embassy, where he made a startling discovery. The treaty, he wrote to Theodore Palmer afterward, "by an accident had been filed without any action, and forgotten!" Hornaday decided that it would be best not to publicize this bureaucratic mishap, and the newly found treaty moved along its way. "Of course you will immediately recognize the necessity of saying nothing whatever about the accident at the embassy," Hornaday advised Palmer. Both the United States and Great Britain signed the treaty in August 1916.[16]

While the treaty would probably have been found eventually if Hornaday had not visited the State Department, it is clear that his characteristic initiative and refusal to wait for results hastened the ratification process. Hornaday continued to ride herd on the migratory bird legislation, enlisting his latest friend, Thornton Burgess, author of the popular children's "Bedtime Story-books" series, to write a book generating sympathy for migratory waterfowl. Published in 1917, *The Adventures of Poor Mrs. Quack* told the traumatic story of a duck couple trying to survive the "Terrible, Terrible Guns." Hornaday could not have written it any better. In gratitude for this and other contributions to the cause, he awarded Burgess the Permanent Wildlife Protection Fund gold medal "for vast services in teaching American children to appreciate and preserve wild birds and animals." The new Migratory Bird Treaty Act, which replaced the Weeks-McLean Act due to the latter's disputed constitutionality, became law in 1918.[17]

Unfortunately, Hornaday's game-refuge bill failed to meet with the same success. Each of the five times it came up for a vote on the floor of the Senate before the session ended in February 1917, a single senator voiced an objection. That senator was Reed Smoot, a Republican from Utah. After Smoot voted "nay" the first time on April 20, 1916, Hornaday sent him clippings from Utah newspapers supporting the sanctuary plan. But this attempt at persuasion failed. As Smoot kept voting against the game-sanctuary plan, Hornaday grew increasingly agitated. In mid-September, the secretary of agriculture informed Hornaday, through the chief of the Biological Survey, that he would not act on the plan until the various state legislatures had passed enabling legislation themselves. In other words, Hornaday's bill stood no chance of going into effect anytime soon even if Congress miraculously passed it. In the end, Hornaday blamed the "abominable 'States Rights' fetish," as he labeled it in a letter to Theodore Roosevelt. He reserved a very cold place in his heart for Smoot. "May curses alight *upon him abundantly*," Hornaday candidly wrote to his half brother David.[18]

The game-sanctuary plan was more than a failed conservation measure. Waged almost single-handedly, it was the most personal of all of Hornaday's wildlife-protection efforts. Although he could take some solace in having inspired local activists, like Aldo Leopold, who successfully lobbied their states to create state game sanctuaries, he never quite recovered from the defeat of his pet project. "As I look back upon the disgusting fate of our campaign for big game sanctuaries in national forests," he wrote in *Thirty Years War for Wild Life*, "I wonder why it did not turn me against conservation causes for all subsequent time."[19]

Personal investment alone, however, did not account for Hornaday's deep disappointment at the game-sanctuary bill. He blamed sportsmen, his former allies, for the defeat of his bill. Hornaday's idea for sanctuaries was to offer wildlife a safe haven for the breeding and repopulation of depleted species, such as mule deer and elk. He hoped the states would adopt sensible hunting regulations so that this effort would not lead only to more killing of game. It was a point he stressed throughout his speaking tour. For this reason, sportsmen, he believed, especially at the national level, did not favor the bill. Though it was not worth their effort to come out openly against it and run the risk of falling victim to Hornaday's poison pen, they withheld their support and let it die of neglect. He explained their motivation, as he saw it, to Aldo Leopold in January 1916, "because they are down on me for advocating so many measures which mean the limitation of shooting." Hornaday concluded that hunters only supported wildlife protection if it could be proven to increase the game they would hunt. "It is a waste of time to rely upon sportsmen to take the drastic measures that are now needed," he wrote.[20]

The game-sanctuary plan marked a watershed in the evolution of Hornaday's conservation ideology. Although he had been moving away from sportsmen for over a decade, the break was now sharp and clear. In January 1917, he openly voiced his discontents in an article entitled "Hunting Ethics" published in *Forum*, in which he sternly denounced sportsmen for opposing sound conservation measures. "Millions of American game birds and mammals have been killed because of the rotten ethics of the hunting field," he thundered. By the end of summer, he had placed T. Gilbert Pearson and the Audubon Society in the same camp as the American Game Protection Association (AGPA), the creation of the gun makers. "Frankly, I am disgusted with Pearson because of his letter to *Forest & Stream*," he wrote to Aldo Leopold on August 24, "and because of his cringing and trimming before the shooting element represented by that paper." In his letter to the editor,

Pearson mildly defended the National Association of Audubon Societies' policies and reminded hunters that his organization was their friend. Based on his past experience with Pearson, especially his waffling over the 1911 subscription fiasco and the failure of the NAAS to support regulations on automatic and pump shotguns, Hornaday interpreted this as a wholesale surrender to the sportsmen. Subsequent events during the war years only served to drive the wedge deeper between Hornaday and both Pearson and sportsmen. In the end, this division shattered the Progressive wildlife conservation movement that had worked so successfully to pass the landmark bills of 1913.²¹

The death of the sanctuary plan coincided with an increase in Hornaday's war-preparedness activity. He was especially concerned about the small size and inexperience of the U.S. armed forces. "Any nation with 5,000 miles of coast line and a fourth-class navy is in a very dangerous position," he wrote in one circular. To remedy the situation, he formed the United States Junior Naval Reserve (USJNR) in October 1916 to train boys aged sixteen to eighteen for eventual service in the U.S. Navy. Cadets trained at Camp John Paul Jones in Corpus Christi, Texas, where they barracked in a local hotel purchased and renovated by the local population specifically for the USJNR. During the summer, cadets went north to Camp Dewey, Connecticut, for more technical training. Plagued by cash problems, scandals at the camp, and the failure to recruit talented staff, Hornaday described his experience at the helm of the Reserve in "Eighty Fascinating Years" as "years of infernal worry and nagging work" and "a personal Waterloo." All his efforts came to naught when Secretary of Navy Josephus Daniels denied the USJNR official recognition because it failed to provide serious professional training to its recruits as required by law. Perhaps the greatest obstacle to his success was Hornaday's own doubts about the entire project. As early as February 1917, he wanted to fold his creation because he believed, as he told Victor Stockell, that during a war the necessity for a complete mobilization of manpower could only be achieved through conscription.²²

The winter of 1916–17 was a deeply troubling time for Hornaday. In early October 1916, Josephine had burned her leg so badly that she was bedridden for a couple of weeks. As always, Will played attentive nursemaid to his ailing wife. When she recovered, they took a vacation to the wilds of New Hampshire, but this offered only a brief and cold respite. The results of the 1916 election deeply disappointed Hornaday. A lifelong Republican, he had hoped his party would capture the White House and establish more muscular defense and foreign policies with a second Theodore Roosevelt presidency. That such an idea was clearly delusional after the Rough Rider's Progressive Party apostasy in 1912 did not deter Hornaday from maintaining that "the jackass leaders of the Republican Party" picked the wrong candidate, Charles Hughes. "Through personal spite they tried to give us the Big Stuff when we demanded the Big Stick," he wrote his half brother David. "They put up a conundrum with whiskers when we asked for a MAN." The result was four more years of Woodrow Wilson and what Hornaday described as the "lack-leadership" of his administration. On top of that, Hornaday's son-in-law, George Fielding, went to Kansas to assist with his father's business, leaving Hornaday, as he admitted to his confidant Edmund Seymour, feeling "depressed" by his absence.[23]

President Wilson's failure to respond forcefully to the German resumption of unrestricted submarine warfare only intensified Hornaday's frustrations. "Do you think that if T. R. [Theodore Roosevelt] were president the Germans would now be sinking our ships like 1, 2, 3?" he asked David rhetorically. "I do *not*." Meanwhile a vocal and erudite group of intellectuals pressed the case for pacifism. Hornaday considered such prominent peace advocates as Jane Addams, Morris Hilquit, David Starr Jordan, Amos Pinchot, Oswald Villard, and Lillian Wald as nothing less than cowardly traitors. To Henry Wood Elliott, he even criticized George McLean, who had cosponsored the revolutionary migratory bird protection bill and generally labored on behalf of wildlife protection, for being "the champion pacifist of the United States." But despite all his rants at the political class, in the end he placed the blame for American isolationism squarely on the shoulders of the great mass of apathetic American people who failed to realize that their interest would best be served by following his course. "But the asinine *majority* of the American people have 'done it to themselves,'" he wrote in contempt. They would be forced to discard their cozy isolationism only when reality struck them hard. He could only hope it happened before the teetering Allies met disaster.[24]

Hornaday's frustration with the American people over the war paralleled his thinking on conservation. He blew his trumpet so loudly in the latter cause because he feared Americans would attain enlightenment about protecting wildlife only after another extinction. It was one more illustration of how the model of the temperance reform movement and his Adventist upbringing influenced Hornaday's thought. To be neutral was to stand in league with the likes of market hunters, the gun trust, the alcohol trust, the Kaiser, and even Satan himself. Only immediate atonement and taking up a sword in the righteousness crusade could avert disaster.

The United States finally entered World War I in April 1917. Hornaday threw himself into the war with his characteristic zeal. "From this time until the close of the war I will be deep in the business of national defense," he wrote in May 1917. "On account of the war, my activities in wild life protection until the close of the war will be much more limited than they otherwise would be." A year later, he had not slackened his pace at all. "I work very nearly all the time that I am awake," he wrote to his half brother David in April 1918, "and Josephine and I have cut off all pleasure trips and festivities of every description. We do not care to enter into any gaieties while good men are being slaughtered in France."[25]

Treating the zoo as if it were itself a state, the director wasted no time putting it on a war footing. "Dr. Hornaday did not singlehandedly win the war," wrote the official historian of the Bronx Zoo in 1974, "but he tried." Hornaday reduced the rations of the animals and replaced chicken and beef in the animals' diets with horsemeat. He conserved fifty tons of coal by moving the administration offices into the birdhouse, and ordered vegetables planted on all available grounds. Vacations for all staff were cut in half. To add a martial flavor, he formed Company A of the Zoological Guards from keepers and curators, and paraded them around the grounds in uniform during the day. They conducted maneuvers for the guests and guarded the park against the increase in crime during the night. An on-site Red Cross station promoted voluntarism. "Every man and woman in the Park has volunteered to contribute each day one hour of extra work for the benefit of preparedness and defense," he proudly wrote to Aldo Leopold on April 23, 1917. "And, believe me, the spirit of self sacrifice is in our midst." The animals survived the war, even if they emerged looking gaunt and slothful. At the end of the war, Hornaday sent more than fifty animals to the war-ravaged Antwerp Zoological Gardens. The following year he donated animals to London. Inflation and constant belt tightening, however, caused some loss of privileges at the zoo. "We are now so economical that I have no messenger, and have not had one for six months," he wrote dejectedly in September 1919. "To think that I should come to this, after twenty years in the Zoological Park! But in view of the treatment accorded to us by the City of New York last year I am thankful that I have an official roof over my head."[26]

Throughout the war, Hornaday strenuously advocated 100 percent loyalty to the United States and cursed all forms of dissent. "The loyal people of America must ruthlessly suppress all disloyalty, expressed, or implied, overt or latent," Hornaday wrote in *Awake! America*, which he authored for the American Defense Society (ADS). Working with the ADS, he waged his first loyalty campaign in the New York City schools in fall 1917. After successfully crusading to ban German language from the curriculum, they went after teachers' political views. At rallies they made public allegations of disloyalty and named individual teachers who declined to sign a loyalty pledge. "A public school teacher should stand in loyalty and Americanism precisely where we expect an officer of the Army or Navy to stand, and should be held to an equally rigid accountability for the slightest symptom of disloyalty or of failure in thorogoing [*sic*] Americanism," a sympathetic letter from Theodore Roosevelt declared. Hornaday considered disloyalty a major threat to the American war effort, and schoolteachers were only the first targets. As he reasoned, without the absolute devotion of his fellow citizens, there could be no total war. And without total war, there could be no victory. "If the Allies fail to win," he ominously warned, "that failure will sign the death warrant of human liberty on this earth." After waiting two solid years for the American people and government to attain the necessary enlightenment to realize that joining the Allies as a belligerent was the only sensible course of action, he was not going to let a lack of enthusiasm or German sympathizers spoil any chance of victory.[27]

Hornaday felt Germany had resigned from the human race and should be punished harshly. Germans "are not reasonable beings," he wrote. They are "amenable to nothing on earth save bayonets and force!" he exclaimed to another correspondent. "I want to see the Huns *exterminated* root and branch," he confessed to his half brother David in May 1918. Hornaday formed the American Guardian Society with the avowed purpose of not purchasing any German-made goods for a quarter century following the war. His personal war against Germany extended to his dealings as director of the New York Zoological Park.

"Personally I am determined never again to buy a wild animal from any inhabitant of the country whose soldiers deliberately smashed up all the machinery of Belgium that they had not already stolen before they retreated from the country, thereby needlessly crippling the industries of an already impoverished nation," he wrote to Dr. Dreyser, the director of the Copenhagen Zoo in Denmark. Bucking the established positions of the day, Hornaday supported both a punitive peace and the League of Nations. In June 1919, he described the American failure to join the league as nothing less than "a calamity."[28]

Hornaday labeled Socialists and Bolsheviks as German sympathizers responsible for increasing the level of misery in an already suffering world. Although the fighting in the war stopped on the eleventh hour of the eleventh day of the eleventh month in 1918, Hornaday, like thousands of other superpatriots in the United States, redirected his vitriol to a new enemy, radicals. He complained to Alfred E. Smith about the presence of undesirables during the victory parade in New York on November 12, 1918. "The City of New York has been disgraced before the nation and the world by the appearance on its streets of a disorderly procession bearing aloft the red flag of anarchy," he wrote. Soon he painted in even broader strokes a picture of a revolutionary communist fifth column growing out of America's urban immigrant communities. "The Bolshevist element is comprised almost wholly of foreigners who are un-American, such as the Russians, Poles, Italians, and others," he charged in October 1919.[29]

Throughout what came to be known as the Red Scare, Hornaday continued to author pamphlets and tracts against the new enemy. His most notable work, *The Lying Lure of Bolshevism*, was published and distributed by the ADS. Bolshevism "means CIVIL WAR, armies, dead men, burned cities, and ruined homes! These may yet be the fruits of invertebrate Americanism, the fatal fetish-worship of the 'free speech' idol and 50 years of dragging American citizens through the slums of all nations," he wrote. Words such as these delighted the journalist and satirist H. L. Mencken, who considered Hornaday's writings simply an "amazing" example of hysteria. Deeply engaged in guarding America against radical immigrant leftists, Hornaday described 1919 as "the most worrisome year of my whole life," so seriously imperiled did he consider the country amid massive strikes and Palmer raids. His actions led some to threaten him, but he took such threats in stride and never considered restraining his tongue or pen.[30]

The war and Red Scare wreaked havoc on the Hornaday household. Now older and more dependent than ever on domestic help, Will and Josie struggled with inflationary prices and poor domestic service. It vexed Hornaday, who always believed that everyone should take pride in their work no matter their position, that some Americans took out their dissatisfactions on their employers or customers. In 1919, he described his domestic help as "a long procession of worthless scarecrows in petticoats posing as cooks." The labor unrest and numerous strikes of the same year personally inconvenienced him. "Don't give way!" he advised the editor of his poetry book *Old Fashioned Verses* on how to handle the recalcitrant workers. "Do not compromise with the enemy." Complicating the inflation, scarcity of goods, and poor help that troubled him throughout the war, the size of his family had increased. But Hornaday greatly enjoyed playing the role of doting grandfather. "You would like his manliness, his self-reliance, his ambition to *do* things," Hornaday wrote of his grandson Temple to his half brother David Miller, "*and* the way he stands up for his rights whenever he thinks they are infringed!" Temple Fielding certainly had the Hornaday blood and displayed some of the same characteristics his grandfather had demonstrated as a child in Iowa nearly sixty years earlier.[31]

In Michigan, Hornaday's older sister Mary became very ill, further adding to his daily worries. Despite the pain that Mary's terminal sickness would have caused him, Mary's sharp tongue created friction with her brother and his wife. "Of course, if Mary had not turned against Josie and me as she did," he wrote to David, "we would have had her living with us long ago and she would be here now, but she has made the situation as *difficult* as possible for me." Yet he traveled to Battle Creek, paid her nursing bills, and bought her medical supplies, while concealing these acts from her. Hornaday continued to help Mary as best he could under these trying conditions until her death in early 1920.[32]

Meanwhile, the cause of conservation had not suffered as badly during the war as many had feared, though with few campaigns, Hornaday reported to Wildlife Protection Fund members that he spent little and invested some of the fund's money in Liberty Bonds. One conservation campaign—and it was a campaign of great significance—pitted Hornaday against elements of the Boone and Crockett Club and the federal government's Bureau of Biological Survey. The importance of this battle lay not so much in the outcome as in the divisive struggle itself, which was a precursor to a larger, more protracted, and even more vicious battle among conservationists through most of the 1920s.

With wartime demands over shipping threatening Alaska (which had to import food) with a genuine crisis, one possible solution was to reduce hunting regulations to allow game meat to supplement the strained food supply. But when Hornaday discovered that this would allow for the sale of game year-round in Alaska, he became incensed. Biological Survey Chief Edwin Nelson and Alaska's congressional delegate Charles Sulzer, the sponsor of the proposed reform of hunting regulations in Congress, "must have been crazy besides, if they thought they could put such a thing over on the men who have been fighting for twenty years to stop the sale of game everywhere," Hornaday wrote to Henry Wood Elliott.[33]

Yet Hornaday unleashed his strongest broadsides on sportsmen who supported the idea, especially Charles Sheldon. Hornaday savagely attacked Sheldon in the *New York Sun* and during House Committee of Territories hearings in February 1918. More stinging still was Hornaday's attack on Sheldon on the floor of the Boone and Crockett Club's annual meeting. Hornaday had been denied membership in the exclusive club in 1894 because he was not a genuine sportsman but something of a mercenary. Grinnell described Hornaday in 1894 to Theodore Roosevelt as "a good man and a pleasant fellow" whose "hunts have been in a measure professional. That is to say, he has been a taxidermist and hired to go out and collect." In 1910, Madison Grant used his influence as Boone and Crockett secretary to admit Hornaday as an associate member, which did not allow him to vote or vouch for new members, but did permit him to attend the annual meetings. Hornaday convinced Charles Davison, secretary of the Boone and Crockett Club, to publicly criticize his erstwhile member in the newspapers, and he successfully lobbied the club to support all the provisions of the proposed reform except the expansion of game-meat sales. George Bird Grinnell wrote a scathing rebuke to Sheldon, who was something of a protégé, but restricted it to a private communication. "It seems to me impossible that you should have thoroughly understood the Sulzer bill," Grinnell wrote in a scolding tone. Sheldon insisted that he understood but argued that unrealistic game laws would only lead to lawlessness and greater flouting of them. Nevertheless, Hornaday's relentless assault left Sheldon so dispirited that he twice attempted to resign from the club. "If Hornaday had not seized the occasion and made it spectacular it could have easily been made all right without noise or the public taking any interest in it," Sheldon wrote to Grinnell. He added that he considered Hornaday "the Bolsheviki element of game conservation." Hornaday's high-handed moralism grated even more on his opponents when he got his facts wrong and refused to recognize misstatements, including a statement before a congressional committee claiming that the 1902 Alaska Game Act that Madison Grant had played such an important role in writing did not allow any sales of game meat. Hornaday's assertion was incorrect, and it made what was in fact a revision of existing policy appear to be a dangerous and radical reversal of current practices.[34]

Hornaday also greatly irritated Edwin Nelson by gruffly complaining directly to the secretary of agriculture about the poor policy of the Biological Survey. "In supporting the Sulzer bill," Hornaday wrote to Secretary David Houston, "your Department has made what I cannot but regard as a most serious mistake." Some of Hornaday's vitriol may have stemmed from the guilty conscience of a man who felt he had failed to be adequately on guard for wildlife as the war had absorbed most of his attention. "I also realize that I have been not only exceedingly stupid in regard to this matter," he admitted to Nelson, "but also asleep on duty in regard to the game of Alaska, and therefore deserve to be shot." Nelson entreated Grant to restrain the New York Zoological Society's director, but Hornaday characteristically firmly stood his ground. "I regard Nelson's mental and moral density and wrongheadedness in regard to the sale of game in Alaska as something very serious," Hornaday wrote to Grant in March 1918, "and it creates an ugly condition that is bound to be troublesome." As for Nelson's complaint that he had bad-mouthed the Biological Survey to the secretary of agriculture, Hornaday announced to Grant that he was "now quite through with apologizing for the Biological Survey." "Whatever may happen hereafter, no one can say that I have not tried my utmost to induce you and Secretary Houston to withdraw from an impossible situation," he wrote to Nelson, intent on drawing the battle lines.[35]

Hornaday successfully fended off the Sulzer bill, although he admitted to Wright Wenrich, an Alaska game warden, that Alaska needed to revise its game laws once the war ended. Nelson spent most of the early 1920s working on legislation to reform the territory's game laws, despite a lack of interest by Congress. Throughout, Hornaday freely and often offered suggestions for the bill. Nelson must have been thunderstruck when he received one such letter in March 1921. "I have thought long and hard on the question of the sale of game meat," Hornaday wrote to Nelson, "and while I hate to admit it, I have come to the conclusion that the situation demands a regulation by which genuine frontiersmen living remote from lines of steam communications should have the right to buy and sell moose meat and caribou meat, but not mountain sheep meat."[36]

Yet the battle over the Sulzer bill put whatever hope remained of reviving Hornaday's relationship with the sportsmen for conservation on life support. His position underwent a significant evolution since he first advocated the sportsmen's values of voluntary self-restraint in *Camp-Fires in the Canadian Rockies* and *Camp-Fires on Desert and Lava*. A series of policy debates over the next two decades, culminating in the Sulzer bill, drove Hornaday and sportsmen increasingly apart until the gap could no longer be bridged. The death of his friend Theodore Roosevelt, the man he once described as "the noblest Roman of them all," in January 1919 saddened Hornaday. The last member of the sportsmen elite he truly respected was now in the grave.[37]

After the failure of his plan for game sanctuaries on federal forests, and amid the bruising fight over the Sulzer bill, Hornaday joined a campaign with Thornton Burgess and *People's Home Journal*, a popular magazine, to encourage individuals to create sanctuaries and preserves on their private land. Admittedly, Hornaday had been cool to the idea when Burgess first broached it with him. In fact, this was before he had enlisted Burgess's help protecting migratory birds, and he had dispensed with the author, whom he did not recognize, rather curtly. But Hornaday had come around to support and publicize Burgess's plan, mostly by supplying awards to people involved from the Permanent Wildlife Protection Fund coffers. Hornaday's interest in the plan was a natural outgrowth of his work at the zoological park to educate his fellow citizens in the natural world, his interest in conservation, and his calls for individual responsibility for wildlife protection. In 1918, the initial year of the campaign drive, more than 2,604 sanctuaries totaling more than 770,000 acres were dedicated, according to Hornaday's account in the Permanent Wildlife Protection Fund Statement. By March 1920, the total number of acres stood at 1,520,000.

In 1922, Hornaday thanked Moody Gates, editor of the *People's Home Journal*, with a gold medal from the Wildlife Protection Fund. The effort concluded in 1923, after 9,000 sanctuaries encompassing nearly 3 million acres had been established. "The majority of the sanctuaries made were in the South, where they were most needed," Hornaday proudly wrote in *Thirty Years War for Wild Life*.[38]

Meanwhile Hornaday fostered individual responsibility for wildlife protection among the younger generation in the Boy Scouts. "The Boy Scouts of today have a solemn duty in the protection of the remaining beasts and birds for the Boy Scouts of tomorrow!" he had written in 1915. At first he had exhorted these young men to educate their peers, report violations of law, and write essays and addresses extolling the benefits of nature in school. The Permanent Wildlife Protection Fund had established a gold medal to award scouts who performed distinguished services in the name of wildlife conservation. But, as Hornaday admitted, the original goals he had set for Boy Scouts were too broad and nearly impossible for young men to achieve, and only one scout and one scoutmaster had earned the gold medal. He had then scaled back the award to a badge and lowered the requirements. Many scouts wrote him directly, and he took the time to write lengthy and encouraging replies. However, as David MacLeod points out, Hornaday failed to win the Boy Scouts of America as an organization over to a more active role in conservation. It remained at the initiative of individual scouts to earn the award, which continues to this day and has been given in Hornaday's name since 1970. More than 1,100 scouts have won the award since its inception.[39]

8

FIGHTING THE ESTABLISHMENT

IN MARCH 1920, Hornaday received what he regarded as payback for his vitriolic attacks on opponents of World War I and others he deemed un-American during the Red Scare. David Hirschfield, the New York City commissioner of accounts, had conducted a year-long, thinly veiled witch hunt into Hornaday's stewardship of the New York Zoological Park that now culminated in a report that dubbed Hornaday "a monarch in his own principality."[1]

Hirschfield's report accused the director of the zoo of deliberately firing a whistle-blower, keeping shoddy financial records, engaging in unethical business practices, and committing nepotism through employing his nephews, the Mitchells, and purchasing grain from his son-in-law's family firm in Kansas. "The whole thing has been a travesty covering the purpose, which has been admitted by some of those interested in his scheme, to get political control of the park and put politicians they want to reward in charge of it," Hornaday responded to reporters who questioned him. To his friend John Phillips, he wrote in private, "Hirschfield being a Polish German Jew is angry at me because of my war on the Germans during the war; and there are others in the Administration who, I have been told, are angry at me because I have made fun in print of the Sinn Feiners," the Irish revolutionaries. Hornaday never seriously fretted over the report. It was more a distraction and "a pitiable object" than a real threat—but it wasted valuable time and strained his relations with city officials during yet another series of protracted negotiations over a wage increase for his staff. Meanwhile the death of his sister Mary the month before and his financial woes in the chaotic postwar economy added to his troubles.[2]

A conflict with the governor of Alaska and sportsmen over regulations in the northernmost territory did not help matters. With the war over and the food supply more secure, Hornaday sided with those who believed in a thoroughgoing reform of Alaska game regulations. But, as was often the case, he argued vehemently with other conservationists over precise details. He especially objected to the contention that brown bears and bald eagles were pests deserving no protection. In February, he issued a pamphlet that argued that it was self-defeating for the game-meat-dependent Alaskans to want the right to kill more game since it would endanger their supply. Despite the relatively small population of Alaska (55,000 according to the 1920 census), Hornaday saw the ominous components (underregulation, construction of the federally funded Alaska Railway, and better firearms) of the buffalo slaughter falling into place once again. The combination of these elements would, according to his reasoning, naturally create the missing part of his formula for extermination, the market for game meat. Governor Riggs found particularly insulting Hornaday's attitude that distant outsiders could better manage the game than locals themselves could do. As he had done with John Burnham almost a decade before, George Bird Grinnell advised Riggs to ignore Hornaday. "He is often abusive and irritating, but I suspect the wise thing to do is to ignore what he says and writes," Grinnell recommended. "Dr. Hornaday represents himself and a proportion of the sentimentalists of the country, most of whom are women and children." Two months later, however, the governor still found Hornaday impossible to ignore. "This man Hornaday, by his lies, is doing more damage to the game than any other person," Riggs wrote of Hornaday's assertion that the changes would lead to the extinction of the brown bear. "He is simply disgusting."[3]

Hornaday's relationship with Grinnell, which had been frosty since the Audubon gun-money debacle nine years earlier, completely broke in autumn 1920. Their positions on hunting had progressed along two separate tracks since the introduction of the automatic and pump shotguns. Over the years, Hornaday had increasingly blamed Grinnell's constituency, sportsmen, for the decline of American wildlife. "Yes, who is really to blame for the absurd hunting license fees, the joke bag 'limits' and the criminally long killing seasons?" he asked in a 1920 edition of the *Wild Life Protection Fund Bulletin*. "The answer is: Up to 90 per cent, *it is the sportsmen themselves!*" Statements such as these and continued reliance on the phrase "game hog" infuriated Grinnell, who considered Hornaday's accusations an ungentlemanly affront to the sportsman's code of self-restraint. But the greater strain in their relationship occurred on a personal level. Grinnell always considered Hornaday's tactics low-class, especially his penchant for turning policy disputes into very bitter personal, public quarrels, and avoided direct confrontation. For nearly twenty years, Grinnell voiced his discontent with Hornaday's wildlife-protection policies to the New York Zoological Society's Madison Grant, arguing that Director Hornaday embarrassed the society. Grinnell believed Grant was the only one capable of exercising a genuine restraining influence on the good doctor. Hornaday found these methods underhanded and maddening. "The cowards do not care to approach me directly," Hornaday commented bitterly in February 1920, "so they take it out in trying to knife me in the back." "I cannot understand the influence that Mr. Grinnell so easily exerts over the mind of Mr. Grant," Hornaday wrote Henry Fairfield Osborn with evident frustration. "All I can do is to accept the fact and make the best of it." But Grinnell persisted, and Hornaday took great offense when he saw Grinnell's correspondence with Riggs published in the Alaska newspapers. "Your letter was a hostile and damaging production," Hornaday seethed in anger to Grinnell. Having received no satisfaction through the mail, Hornaday marched over to Grinnell's house and confronted him directly. Although Grinnell expressed "regret" that the director had "been annoyed," he refused to recant his comments, as Hornaday reported to Marshall McLean.[4]

The Alaska debacle became even more divisive in the territory and more personal for Hornaday when he openly endorsed a candidate, Daniel Sutherland, for delegate to Congress. In moments like this, when Hornaday felt he was championing the good fight, his self-righteousness inflated his ego. Like the Audubon board's decision to accept the gun maker's money in 1911, Hornaday placed himself in the forefront of the minds of the people involved. "The campaign has been merely a series of attacks upon me," Hornaday wrote to Biological Survey Chief Edwin Nelson, "with an incidental campaign to beat Mr. Sutherland who was running for delegate from Alaska." Hornaday at least enjoyed the satisfaction of Sutherland's victory, something that gave him pause for positive reflection as he lay prostrate after a painful eye operation in November 1920. Republican Warren Harding's landslide victory in the presidential polls also benefited Hornaday. The territorial governorship was a patronage appointment, and as soon as the administration took office, it removed the Democrat Riggs from office and replaced him with someone friendlier toward brown bears.[5]

The fight over Alaska game regulations was just the opening skirmish of the larger and more ferocious war among wildlife protectionists that tore them apart and ultimately destroyed the Progressive wildlife conservation movement. Ironically, the source of the destruction flowed directly out of the movement's greatest success, the establishment of the Migratory Bird Treaty Act in 1918. Now, three years later in 1921, Senator Harry New of Indiana and Representative Daniel Anthony of Kansas, both Republicans, responded to Edwin Nelson's repeated calls for waterfowl refuges and introduced a migratory bird refuge bill into their respective chambers of Congress. No one would have guessed that the principles of the legislation would so devastatingly fracture the movement.[6]

As Nelson was calling for waterfowl refuges, John Burnham and the American Game Protection Association (AGPA) created by the gun makers were making a case that the combination of the Weeks-McLean Act and a great reduction of hunting during the war had caused such a dramatic growth in the waterfowl population that hunters could enjoy the full fruits of their conservation for the first time. No hard science substantiated this thesis of a massive population increase, but it seemed logical enough, and sportsmen would stubbornly cling to it throughout the remainder of the 1920s and into the 1930s. Burnham's answer to the question of how to meet the needs of hunters while creating waterfowl refuges was to demand public shooting grounds on the safe havens. To further tie the existence of the refuges to the interests of sportsmen, he suggested a federal hunting license to pay for the land acquisition. Nelson voiced no objections and these particulars were written into the New-Anthony bill.

William Temple Hornaday expressed uncharacteristic ambivalence toward such important legislation as the New-Anthony bill and did not think it could pass the Congress. He was, as he wrote to his friend Edmund Seymour, "not at all sanguine about the ultimate success of the measure," blaming his old nemesis, the states' rights legislators. His prediction proved prescient, as the bill passed the Senate, but failed in the House.[7]

Throughout 1921, Hornaday worked on *The Minds and Manners of Wild Animals*. Charles Scribner's Sons published the book, and it appeared serially in the Sunday edition of the *Washington Post* in 1922. He dedicated the book to his associates of twenty-five years, the "officers and men" of the New York Zoological Park. Years of studying wildlife in both the wild and controlled environments had convinced Hornaday that animals possessed the ability to reason and learn. He believed that the truth about their reasoning and learning lay somewhere between the nature fakers who ascribed unnatural skills to animals and those who considered animals to act solely on instinct without any thought. "I entirely dissent from Mr. John Burroughs' views that wild animals, and for that matter, other animals, know what they know only by instinct and inheritance," he had written two years before the book was published. Hornaday's notion of animals possessing reasoning abilities was sound enough, but not revolutionary. His arguments echoed those first made by George Romanes in *Animal Intelligence,* a work already more than thirty years old. But Hornaday's arguments aided his conservation efforts in that they supported the idea that animals shared some emotions in common with humans and therefore ought to be valued as more than just economic resources or objects of sport. Frank Chapman commented favorably in the Audubon Society's *Bird-Lore* magazine on Hornaday's ability to create sympathy and even feelings of kinship for animals, which Chapman considered no small feat.[8]

As a purely scientific treatise, the book was clearly lacking, however, and did not deserve Hornaday's own assessment later in the decade that it remedied the ill-effects of the nature fakers and "wrought in the United States a stabilizing effect on the public understanding of mammals, birds, and reptiles." Instead, it was a highly impressionistic and anecdotal work that dismissed the prevalent intelligence-measuring methods of the day in favor of nondescript and vague tests that could not be re-created or verified. Hornaday's conclusions essentially resulted from a lifetime of observation. He concluded that the chimpanzee was the smartest animal, followed by the domestic dog, domestic horse, Indian elephant, and orangutan in a tie for second place. The rhinoceros had the dubious distinction of being the dumbest member of the animal kingdom. Hornaday's views of animals rested on a highly subjective superstructure that resonated so emphatically in *The Minds and Manners of Wild Animals* that Carl Van Doren of the *Nation* commented, "He cannot be content to keep within the evidence." *Minds and Manners of Wild Animals* was larded with the junk science of its day, like dividing the wildlife population into phrenology classifications, and Hornaday's own anthropomorphic language, such as labeling wolves "criminal," which jeopardized any attempt to provide an unbiased assessment of animals' mental capacities. "The book should be taken, in the opinion of the reviewer, as a notable contribution to natural history, not as a scientific treatise on comparative psychology," wrote the psychologist Robert M. Yerkes in *Science* magazine.⁹

On December 12, 1923, Hornaday finally attended his first meeting of the Migratory Bird Advisory Board, the panel that had earlier been set up to advise Secretary of Agriculture David Houston on bird-protection regulations, in Washington, D.C. He had intended to attend the meeting in December 1922, but Josie's unexpected illness had kept him at her side. Instead, Marshall McLean, a friend, had introduced a motion to limit the use of shotguns, but it was summarily voted down, causing Hornaday to comment to the board's chairman, John Burnham, with typical bluntness, "It seems to me that beyond a possibility of doubt the sportsmen of the United States are going to exterminate their game and their sport." Now, a year later, Hornaday proposed a reduction in bag limits, the legal number of ducks and geese a hunter could kill in a day. As where Burnham calculated that waterfowl were "in the midst of a hopeful and vigorous renaissance," as he phrased it, Hornaday continued to apply his Malthusian formula that humans, aided by ever-improving technology, killed wildlife far in excess of its ability to reproduce. In the 1920s, Hornaday expanded the circle of those who profited beyond the immediate circumference of firearm and ammunition makers to include outfitters, camping equipment firms, campsite owners, guides, and a plethora of other associated businesses incidental to hunting, including automobiles. In the end, he concluded that if he could not restrain the numbers of hunters or the guns used—although he still had not completely surrendered the notion of obtaining some limitation on the use of pump and automatic shotguns—then he would advocate a new measure to restrict the number of birds that a single hunter could kill in one day. For the rest of his life he would hone in on bag limits as the remedy to America's vanishing game.¹⁰

Predictably, Chairman Burnham scoffed at the idea of bag-limit reductions, considering them not only unnecessary, but also a misguided, if not dangerous, folly. "Our American system has been challenged," he said in a long harangue against the proposal, "the system that aims to preserve for all the people the benefits that otherwise will accrue only to those blessed with considerably more than an average share of wealth." Burnham's Rooseveltian rhetoric carried the day, and only Edward Howe Forbush of the board's eighteen members joined Hornaday in voting for reduced bag limits. In mirror image to Hornaday, the sportsmen resisted all efforts to cut bag limits, and regarded his obsession with them as nothing more than the idée fixe of an unstable mind.[11]

In private correspondence, Burnham attacked Hornaday's masculinity, as Grinnell had done earlier. "The war produced an effect on the women of the country which Doctor Hornaday has skillfully guided into a resentment against killing for sport any living creatures," he wrote six months after the vote. But for all of Hornaday's inability to compromise with sportsmen, it must also be recognized that their image of him as a bomb-throwing radical of feminist values, a sort of Carrie Nation of conservation, contributed greatly to their own failure to understand the changing times or to fully comprehend the doctor's appeal to a larger, nonhunting audience. At other times they referred to Hornaday's supporters as "zoophiles," another code for sentimental, nonhunting animal lovers. This was yet another sign of how sportsmen failed to grasp Hornaday's appeal to the larger public, the effect of over 1 million visitors to the zoo each year, and, how, more significantly, game came to be redefined more broadly as wildlife.[12]

Hornaday saw himself quite differently. To him the problem was the existence of what he dubbed the "Eastern Establishment," which was based in New York City and Washington, D.C. He had long recognized the influence of the East on the wildlife-protection cause. When raising money for the Flathead Bison Refuge, he noted with some surprise that urbanites in the East supplied most of the money. Nearly a decade later, in 1917, he reckoned that 80 percent of the work done on behalf of wildlife was done in New York City alone. It took the collusion of the Bureau of Biological Survey and sportsmen in the debate over regulations in Alaska for him to come to see this as a purely negative force. As he conceived it, the conservation associations, leading scientific organizations, government agencies, and gun makers constituted, through an interlocking directorate, a trust that sought to subvert wildlife-protection policy for its own financial gain. As a prominent sportsman, president of the AGPA, and chair of the Migratory Bird Advisory Board, John Burnham was the mastermind of this insidious organization, and George Bird Grinnell was its chief counsel. T. Gilbert Pearson of the NAAS was its advertising director. Although the southerner did not fit the profile of the others with their New York and New England upper-crust credentials, he failed to challenge them, and, in fact, supported them and their dangerous proposals uncritically (at least to Hornaday's mind). Hornaday's litmus test was a provision that would place public shooting grounds on any refuge created by the federal government. He considered it not merely poor policy, but as inherently evil and a sign of the establishment's descent from high ideals to lowly profit. Worse, it would be a path to extinction and extermination. Once again, he was battling for the heart and soul of conservation and would give no quarter to his enemies. As Hornaday saw it, he fought on behalf of the millions who desired some form of wildlife protection, but did not want to see their government become the agents of self-interested hunters.

That night, after losing his bag-limit vote, Hornaday dined with a new friend, Will Dilg, a gaunt Chicago marketing executive and founder of the Izaak Walton League (IWL) in 1922. Named after a famous Scottish angler and author of the seventeenth century, the IWL provided a grassroots conservation advocacy organization throughout the Midwest, far from the gaze and influence of the Eastern Establishment. Dilg's view of sportsmen matched Hornaday's own. "My appeal is not to save the game and fish for sportsmen," Dilg wrote in the IWL newsletter. "I do not care anything about saving game and fish for sportsmen. I want to save something of vanishing America. For its own sake!" Dilg's signature piece of conservation was passage of the Upper Mississippi Refuge Act of 1924. This legislation created a 300-mile wildlife refuge along the shores of the Mississippi River running from Wabasha, Minnesota, to Rock Island, Illinois. In a move more audacious than the refuge's sheer size and length, Dilg demanded that the federal government foot the $1.5 million price tag for acquisition out of the general treasury. And to top it all off, Dilg succeeded against the entrenched power of electric-power baron Samuel Insull, who favored drainage (which would eliminate the wetlands vital for the refuge) because it was good for his business. Hornaday found such boldness exhilarating when compared to what he considered the stale, timid orthodoxy enforced by the interlocking directorate of Eastern Establishment conservationist organizations based in New York. Dilg inspired Hornaday with a model to put forth an alternative to Burnham's migratory bird refuge plan. Hornaday would present the boldest proposal for protecting wildlife, separate the interests of sportsmen from refuge acquisition by advocating funding from the general treasury instead of user fees, and take on the sportsmen and Bureau of Biological Survey in the same manner that Dilg defeated Insull.[13]

After his meeting with Dilg, an emboldened Hornaday applied direct pressure on the Biological Survey's Edwin Nelson to reduce bag limits. "This is of enormous importance," Hornaday wrote in the opening paragraph of a letter to Nelson, "because you are now the last remaining obstacle that blocks the road to a great reform." His aggressive efforts to convince the Agriculture Department of the error of their ways created a backlash in the Bronx. Henry Fairfield Osborn, now president of the New York Zoological Society, ordered his zoo director to stand down from insulting government officials. "Your injunction to me to stop my campaign and keep silent, is mighty alarming," Hornaday replied. "Up to date there is *no fight* anywhere,—merely a difference of opinion between gentlemen." Hornaday remarked to Madison Grant that it was a "physical impossibility" for him to keep quiet on such a monumental issue, and "I hope that President Osborn is not going to insist on the literal execution of his very drastic order to me." Hornaday refused to relent, and the cycle of his offensive letters followed by complaints to his bosses continued throughout the summer and fall. In a letter to Dilg marked "confidential," Hornaday disclosed, "believe me, at one time they came mighty near making it necessary for me to resign from the Society in order to go on!" On a more personal note, he indicated that the intensity of the campaign to reduce bag limits left him exhausted. "This campaign has taken a great deal out of me," he wrote, "and I am feeling very tired. I do not seem to ever get really rested." But there could be no doubt that he would soldier on.[14]

Having no luck with the Agriculture Department, Hornaday sought to lower the bag limit through legislative action. Just as he was polishing his bill, Biological Survey Chief Nelson invited him to the Waldorf Astoria Hotel in New York City for a meeting on December 7, 1924, just over a week before the Migratory Bird Advisory Board would convene again. Although he told his friend and ally Captain A. E. Burgduff, the Oregon state game commissioner, that he had accepted the invitation "with alacrity and hopefulness," his response to Nelson was far from gleeful. "While I have little hope that we can agree on anything regarding the killing of migratory game," he told Nelson, "it can at least do no harm to talk the matter over." Nelson, the nation's head game protector, however, had a surprise for Hornaday. "Now the fact is I believe that the bag limits on migratory game should be reduced," the chief of the Biological Survey said, according to Hornaday's account. "John Burnham thinks so, too." Nelson asked only for time to circulate a brief questionnaire so that he could determine the best number for the bag limit. Hornaday described the meeting as "tremendously satisfactory."[15]

In the early days of 1925, Nelson sent out more than three thousand surveys to game commissioners, wardens, and others who worked daily with waterfowl. As the results trickled back to Washington, Burnham's migratory bird refuge bill, the New-Anthony bill, funded by a hunting license and containing a provision for a public shooting grounds, expired in Congress without having come to a vote. Disappointed supporters planned to reintroduce the legislation in March after the new Congress was sworn in.

Hornaday waited impatiently for the results of Nelson's survey. He was gravely disappointed to receive a letter from Burnham on May 10, with an enclosure from Nelson stating that, alas, there would be no reduction in the bag limits. Hornaday indignantly made his official objection to Burnham, the chairman of the Migratory Bird Advisory Board, gamely adding: "Really, as one campaigner to another, I must congratulate you upon the completeness of your success in holding the Department of Agriculture in line with your views and desires ever since December, 1923." "Now that the bill is being used as a stalking-horse for the benefit of the gun and ammunition people on the one hand," Hornaday wrote Dilg after learning of the announcement, "and the Biological Survey on the other, I am no longer its friend or supporter." He was less flippant to Madison Grant. "The high-brow conservationists of the East have acted outrageously," he wrote his boss, "and they alone are chargeable with the present attitude of the Department of Agriculture, which, by the way, is going to breed trouble of several kinds."[16]

In July, Hornaday went to Spokane, Washington, to speak at a convention of western state game commissioners. He traveled with Will Dilg, and the two allies labored to win over western game commissioners, whom they viewed as a natural counterbalance to the Eastern Establishment. During the journey by rail, Hornaday conducted a bird survey from the train and enlisted fellow passengers to assist in the endeavor. "Throughout the whole journey west of Indiana," he wrote in astonishment, "water was plentiful, and we should have seen at least 10,000 ducks. Instead of that, we saw all told, even including the many lakes of the Yellowstone Park, *precisely 28 ducks!*" Although hardly a scientific study, it supported his contention that bird populations were declining, not thriving as the Migratory Bird Advisory Board and Biological Survey presumed.[17]

Later in the month, as Hornaday was digesting the results of his informal, high-speed study, Nelson issued a formal report claiming there had been anywhere from a 10 to 200 percent increase in ducks and geese, depending on the locality. Nelson reiterated his long-standing complaint on the destructiveness of drainage for agricultural and urban development, but did not feel it necessary to change the waterfowl bag limit. On the contrary, he contended that reduced bag limits would only encourage lawlessness among hunters, a common line used by sportsmen who considered it a poor conservation tool.[18]

In light of what happened, it is odd that Hornaday chose in *Thirty Years War for Wild Life* to deny that the report ever existed. "They were stacked up in Dr. Nelson's office," Hornaday wrote of responses to Nelson's survey, "and from that day to this no report whatsoever of the results of that 'investigation' ever has appeared!" Adding to the peculiarity of Hornaday's historical amnesia, he had even written a typically incendiary press release that the *New York Times* turned into a front-page story. "The Federal Government is in close agreement with the paid agents of the big corporations who make and sell guns and ammunition," he fulminated. He alleged that the root problem remained the connection between the gun makers, through John Burnham and the Advisory Board, to the policy making at the Agriculture Department. Hornaday's words seemed too pointed for some. "Dr. Hornaday is undoubtedly sincere," the *Nashville Banner* wrote, "but he spoils whatever case he has by aspersing the motives of the Biological Survey." Hornaday's words would come back to haunt him.[19]

At the end of August, a convention of state and provincial game commissioners and game conservation organizations convened in

Denver, Colorado. It was a contentious meeting as two bitterly divided factions met to discuss the migratory bird refuge bill. "A gleaming spear was poised at every tepee entrance," T. Gilbert Pearson wrote of the Denver meeting. Hornaday chose not to attend. He had plenty of his younger partisans present to do his bidding, foremost among them Will Dilg. In the end, David Madsen, Utah game commissioner, and Pearson forged a compromise refuge bill that removed the hunting license and replaced it with options for either a tax on cartridges or general treasury funding. Hornaday commended this compromise, but questioned its gravity, and wondered if Dilg really thought that Burnham was truly abandoning his own bill. Regardless of the agreement, Hornaday grew increasingly convinced that it was absolutely necessary to reduce bag limits; otherwise, the refuges, ringed with public shooting grounds, would serve no purpose.[20]

Hornaday's skepticism toward the Madsen compromise proved correct, as the AGPA and NAAS backed away from the Denver convention's handiwork almost as soon as the ink had dried. In *Adventures in Bird Protection,* Pearson defended the reversal of the NAAS, claiming that numerous senators and representatives advised him that they would not pass an appropriation for refuge acquisition out of the general treasury. Burnham, however, never had his heart in the compromise. He sensed that the duo of Dilg and Hornaday had, as he described it, "laid a trap" in Denver, and he hastily extricated himself from it as soon as he could. "John Burnham's influence for evil is perfectly amazing," Hornaday commented to Dilg on the reversal, but he did not spare Pearson. "T. Gilbert Pearson thinks and acts exactly as Burnham does," he wrote another correspondent, "and as Burnham wishes, and tells him." Pearson heard some similar criticism directly from Audubon state directors uncomfortable with the national association for cozying up too close to the sportsmen and supporting their pet public-shooting-grounds provision, but he simply dismissed it. If nothing else, it proved Hornaday's contention that a small number of powerful men in the East pushed a bill that did not enjoy the support of the entire conservation community, especially once all its particulars were known.[21]

Hornaday's combative attitudes toward other conservationists led one person to think the director, who was now seventy, was losing his mind in his old age. "Hornaday is playing one of those rather shrewd games in his subtle appeal to the public at large against the gunners that is very often characteristic of senility," William C. Adams, a Massachusetts conservationist wrote. George Grinnell collected these and like comments regarding Hornaday and passed them to Madison Grant. "I have heard lately of half a dozen people who have resigned from the N. Y. Zoological Society because of the antics of Dr. Hornaday," Grinnell relayed to Pearson. Ailed by crippling arthritis that confined him to a wheelchair and often seeking comfort in spas and sanitariums, Grant nevertheless acted sternly. "I must do *nothing* that will make your situation painful," Hornaday replied to one stricture from Grant, "and nothing that will injure the Zoological Society." But he worried how he could extricate himself from the campaign without, as he described it, "being held up to public scorn as a slacker and a quitter." It appears that a deal was worked out whereby this would be Hornaday's last campaign, although subsequent events would demonstrate that there was a lack of clarity as to what "campaign" they agreed would be the "final" one.[22]

Advocates of the New-Anthony migratory bird refuge bill remained convinced that they could contain Hornaday and see the bill through to success. They were as blind as Hornaday in failing to take into account the toll that the internecine war among the leading conservationists was taking on the morale of the entire wildlife-protection movement. "Between their fights and Hornaday," Robert Sterling Yard wrote of the situation in the spring of 1926, "every ranking game organization has lost public influence which it cannot regain in years." Gustavus Pope expressed the attitude of many sportsmen when he wrote, "I am disturbed and I am frank to say I am disgusted by the continued conflict between these groups, and I am sure my own position is that of a great many sportsmen throughout the country." Men like Pope and Yard were caught in the middle of what was a very personal battle between two strong-willed men over substantive details. It followed the familiar pattern of discord between utilitarian and preservationist conservationists. Pledged to represent sportsmen (and the gun makers), the utilitarian Burnham could not accept a bill without a public shooting ground provision. The refuges were pointless to him if they could not be used for hunting. The self-appointed spokesperson for the millions who did not hunt, the preservationist Hornaday could not accept a bill with such an item unless there was a corresponding regulation to offset its effect on wildlife populations. To him the purpose of the refuge was to propagate wildlife, not just game. The Popes and

Yards of the nation, who believed strongly that a migratory bird refuge was vital and considered the other issues secondary, had little voice in this struggle. Yard was correct, however, in his analysis. The battle between Burnham and Hornaday caused the wildlife-protection movement to stagnate over a single issue. In searching for a way to break the logjam, Hornaday offered an explosive bill.[23]

In mid-March 1926, Senator Royal Copeland, a New York Democrat, and Congressman Schuyler Merritt, a Republican representing Hornaday's home district in Connecticut, introduced a bill to substantially reduce the bag limit on migratory waterfowl. The proposal would cut the bag limit for ducks, for example, from twenty-five to fifteen. Although it stood no chance of success, its sheer audacity stunned the conservative conservationists. Hornaday's bill would reverse the policy established under the Weeks-McLean Act and transfer the power to regulate bag limits from expert government scientists to the uneducated and more malleable members of Congress. Grinnell asked Merritt why a levelheaded man would introduce such a bill. "It seemed to my untutored mind that, irrespective of whether the theory of the bill was right or not," the congressman replied, "the bag limits proposed were about as much as any ordinary man ought to want to shoot in one day." But if Merritt saw the sense of the bill, sportsmen viewed it as yet another sign that Hornaday wanted to eliminate hunting altogether. "If the ducks are decreasing at the alarming rate which Hornaday imagines in his fevered brain," one sportsman wrote to Grinnell in March 1926, "then nothing short of [the] abolition of shooting could possibly save them."[24]

At this opportune moment, John Burnham picked up his pen and dashed off yet another complaint to Grant. "I have no bone whatever to pick with Dr. Hornaday in his campaign for reduced bag limits," Burnham wrote to Grant, "but simply with his libelous attacks." "But if he is an agent of yours in one case, he is also in the other." The following month Grant ordered Hornaday not to appear at hearings on the bag-limit bill. Hornaday protested to no avail. The hearings never took place.[25]

The fighting intensified on April 29, 1926, when New York congressman Fiorello LaGuardia attacked the Burnham bill including the public-shooting-grounds proposal as a sham "under the guise of conservation." His denunciatory speech would probably have gone unnoticed if he had not read verbatim documents of internal AGPA discussions on the floor and into the *Congressional Record*. Already smarting from Hornaday's bag-limit proposal, AGPA leader and chairman of the Federal Migratory Advisory Board Burnham described LaGuardia's performance as a "rotten attack." But the damage was severe, as the details on the AGPA discussions were embarrassing. In one letter, written in 1924, Burnham had stated, "The sentimentalists led by Doctor Hornaday are demanding cuts in the bag limits and seasons, which if carried to the logical conclusion means the reduction of shooting opportunities to the vanishing point. Of course, if this happens, the sale of firearms and ammunition will be seriously affected." The spirit of this communication confirmed all of Hornaday's worst fears and public assertions regarding the deadly nexus of gun-company money and conservation. Edward Howe Forbush, the Massachusetts state ornithologist who was a long-standing friend and ally of Hornaday and an influential Audubon member, considered such clumsy and impolitic comments fatal to the success of the bill. "It can never be passed in the world if the material is widely distributed," he wrote. It was alleged that these AGPA documents came from Hornaday, probably because of the anti-Burnham invective and LaGuardia's friendly disposition toward amending the bill to include a mandatory bag-limit reduction, but he denied it, describing the congressman's speech as "a spontaneous production."[26]

On May 20, 1926, at the annual meeting of the Board of Managers of the New York Zoological Society, William T. Hornaday announced his retirement, effective on June 1, after thirty years of service as director of the New York Zoological Park. At the time, it was rumored that Madison Grant and Henry Fairfield Osborn had forced him out because he had continuously refused to comply with their cease-and-desist orders regarding the bag-limit bill and his vitriolic comments toward federal policy makers. Frederick Walcott, a member of the New York Zoological Society Board of Managers, declared that Hornaday "paid the extreme penalty" for his constant attacks on the Biological Survey generally and for the King bag-limit reduction amendment to the migratory bird refuge bill specifically. Although he had written a chapter for Hornaday's *Wild Life Conservation in Theory and Practice*, Walcott came to believe his former coauthor "has been an exceedingly valuable man, but he now has the mania of persecution and his state of mind, which I presume governs his actions, is pathologic." Hornaday's constant cries of extinction, like the boy who cried wolf, his unremitting attack on John Burnham and Edwin Nelson, and his bag-limit proposals convinced Walcott and many others that Hornaday was mentally unsound. There is more than a hint of truth in Walcott's description of an aged and fixed mind incapable of adjustment. Hornaday had become convinced that waterfowl were declining, fixated on stopping the implementation of the public-shooting-grounds provision, and highly critical of Burnham's connection to the gun companies and Nelson's reliance on Burnham. However, it was uncharitable to suggest that these were characteristics of insanity or that his concerns were totally unjustified. John Burnham did work for the gun companies, and he advised the government on migratory bird regulations. Extinctions, as the buffalo and passenger pigeon proved, could come rapidly. That was a lesson he had learned too well. Clearly, men like Walcott no longer considered Hornaday stable enough to manage the great affairs at the New York Zoological Park. But Hornaday had intimated to Louis Agassiz Fuertes in January, five months before, that "I am soon going to retire from active life." As he was now seventy-one, it should not have been a surprise that he did so.[27]

Whether or not pressure hastened his departure, there can be no denying Hornaday's enormous impact on the New York Zoological Park and the 44 million individuals who passed through the gates during his reign. "What youthful and adult New York owes Dr. Hornaday for innocent and useful entertainment cannot be written in words," commented the *Times*.[28]

Hornaday did more than delight millions of New Yorkers. He revolutionized the concept, meaning, and style of zoos, transforming the stale animal displays of the nineteenth century into the modern educational model familiar today. Furthermore, he used his powerful influence to shape other zoos in the United States. He advised directors frequently both formally and informally, and on several occasions traveled to other cities for prolonged consultations. In the process, he promoted a more preservationist and sympathetic attitude toward wildlife. Americans began to consider wildlife a national, natural treasure to be appreciated, not simply as an economic resource to be exploited. Hornaday's zoo surely played a role in this transformation. That was an enormous accomplishment.

As a lasting achievement he built a solid foundation of infrastructure and staff. W. Reid Blair, the park veterinarian whom Hornaday described as "the best of all men for the place," succeeded him as director. The first-rate team of curators he had installed remained in place and continued to seek guidance from the mentor in his retirement. Blair continued many of Hornaday's policies, but expanded the scientific focus by bringing in a biologist to serve as curator of education. The stock market crash and Great Depression precluded any significant alterations in the design Hornaday envisioned in 1896. It was not until the late 1960s that more meaningful changes were made to the Bronx Zoo.[29]

On June 10, the society hosted a retirement party in Hornaday's honor and presented him with a gold medal. No one said anything about his attacks on the Biological Survey, the migratory bird refuge bill, or bag limits. "I think that retirement was one of the most popular acts of my somewhat checkered life!" Hornaday wrote to his friend Edmund Seymour later. Despite the criticism he had received, Hornaday remained steadfastly loyal to the society and never hinted at any wounded feelings. To the contrary, he embellished the record of his relationship with his bosses at the society. "Throughout the past 33 years the writer of this volume is unable to recall even one wild life protection cause which the Zoological Society failed to support because of doubt about the consequence of such actions," he wrote in *Thirty Years War for Wild Life*. "In those matters, President Henry Fairfield Osborn, and Secretary and President Madison Grant never took counsel of their fears." Putting the best face forward for the public, as he did in *Two Years in the Jungle*, he never wrote publicly of the cease-and-desist orders he had received from his employers, the continuous back-channel efforts of Grinnell and others to restrain his freedom of action, or the lukewarm attitude of his bosses toward his most prized conservation objectives.[30]

Retirement did not represent a clean break from the society. Hornaday remained actively involved with the Zoological Park to some degree up until his death. He completed the last edition of the *Popular Official Guide* he had been working on at the moment of his retirement, dispensed advice via mail, met with his old hands several times a year, returned to the park for ceremonial and special events, and contributed to publications as needed. In November 1928, Madison Grant asked Hornaday to write the official history of the New York Zoological Society. "As to the history of the Society," Grant wrote, "I finally decided that the whole matter had better be written up by you in book form, covering the entire period." Hornaday's successor, W. Reid

Blair, relied on his advice regularly, as did the curators. In keeping with the continuing association between Hornaday and the society, the Board of Directors voted him the title of director emeritus with a generous pension.[31]

The work of transferring responsibilities to Blair precluded Hornaday from traveling to Iowa in June for the dedication of a tablet in his honor at the Ames campus. Hornaday now stood in the upper ranks of Iowa's favorite sons, with cartoonist Jay "Ding" Darling, author Hamlin Garland, and Secretary of Commerce Herbert Hoover. Acknowledging Hornaday's stature, his alma mater at Ames commemorated the place where he had discovered his calling in taxidermy with a bronze plaque. "This tablet commemorates the work of Dr. W. T. Hornaday for his contributions to Zoology and conservation which have been of immeasurable benefit to America. 'It was on this campus as a student, June 1873, that I found myself,'" it read. "For once in my life, I really don't know what to say!" Hornaday replied when he read the transcription. After the bruising months of bitter and personal attacks, it was an enormous consolation.[32]

Following his retirement, Hornaday took the time to work on a series of articles entitled *Wild Animal Interviews,* a collection of fictional conversations between Hornaday and animals published by Charles Scribner's Sons and serialized in the *Evening World*. The "conversations" contained his usual mixture of humor, morality, and conservation philosophy, such as when he advised a crocodile, "The voting power of the hunter and game-killer is too strong." "My folks think these stories are the best things about animals that I have written," Hornaday remarked to a friend, "and that is also the idea of yours." Scribner's, however, had not been as certain at first. Such a departure in format had alarmed his publisher, who requested significant alterations to the first draft. After reading the revisions, Hornaday's editor commented, "It seems to me that you have done as well as possible with the material." Despite this rather odd deviation from natural history, it fit well into Hornaday's practice of presenting animals in a more favorable light. "It is extremely sympathetic and should be very valuable in the interests of wildlife conservation," Madison Grant wrote.[33]

If Hornaday's opponents thought that his retirement would weaken his voice in the field of conservation, they were sadly mistaken. "There was no one available to take over my work and drive it forward to overtake our enemies and if possible head them off from some of their objectives," he wrote in "Eighty Fascinating Years." "I had to carry on!" If anything, retirement only gave him more time to work on the pamphlets and newspapers that he financed through the Permanent Wildlife Protection Fund. Every morning his car picked him up and drove him to an office he maintained in Stamford. "My office is calm, quiet, well lighted, *free* from telephones and interruptions, and the company is fairly genteel," he wrote in an invitation to his old friend Charles Davenport.[34]

Hornaday took an interest in local affairs, accepting a seat on the Shade Tree Commission, advising on Stamford's parks, and serving as honorary president of Stamford's natural history museum. He could not escape his prominence, and several local boys came to him as moths to a light. Foremost among them were John Ripley Forbes and Donald Shipley, themselves to enter scientific and museum fields when adults. "He always had time to talk to anyone on the subject of the wildlife he loved so well," wrote Shipley, who was eight years old when Hornaday retired.[35]

In December 1926, some six months after Hornaday's retirement, the Biological Survey's Edwin Nelson softened his position on bag-limit reductions and reconsidered their necessity. "The facts which I gathered during the last two years have convinced me that such a time has arrived," he wrote to the president of the Currituck Shooting Club. In a stunning reversal, the Migratory Bird Advisory Board voted sixteen to two in favor of recommending to the secretary of agriculture a daily reduction in bag limits from twenty-five to fifteen for ducks and from eight to two for geese. If nothing else, this pendulum swing of groupthink proved Hornaday's case that sportsmen, who had a prominent voice on the board, and the Biological Survey thought as one. But the board's reversal caused such an outcry among hunters that Secretary of Agriculture William Jardine took an unprecedented step and called special hearings for January.[36]

After Nelson digested the outcry and other information he had gathered, he once more changed his position. He decided in March that bag limits would have little effect, as many states already had on the books, or appeared on the verge of implementing, similar reductions. "It is plain, therefore, that a reduction of the Federal limit to 15 a day would fail to reduce the killing in any way in such States," he asserted. Instead, Nelson recommended a shorter hunting season. "A strange and serious occurrence has just taken place in the federal Department of Agriculture," Hornaday wrote in the opening of a Permanent Wildlife Protection Fund press release. "Some very powerful influence, evidently from without, has caused Dr. E. W. Nelson to suddenly abandon the excellent intentions announced by him on Jan. 17 and 20, and also later, regarding the amelioration of the hard conditions now operating against the ducks and geese of the United States."[37]

At the end of April, Hornaday left the bag-limit debate for a short trip to Iowa. He wanted to see his commemorative tablet, and the college had asked him to address the student body. As one of its most famous alumni, Hornaday received the red-carpet treatment, including an honorary banquet. He gave one convocation address on the conservation of wildlife and another for the entire student body entitled "A Slump in American Morale." In this latter speech, Hornaday uncorked his decade-long angst on the decline of American culture. On the positive side, he advised his captive audience to find "an occupation which will furnish maximum interest and satisfaction" and warned them not to become solely consumed by the quest for money.[38]

Three months after his return from Iowa, the conservation battle got nastier when John Burnham sued Hornaday for libel and $2,500 in damages. The lawsuit focused on Hornaday's comments that asserted Burnham was a dupe of the gun makers that were published in the *New York Times* in August 1925. Burnham, the chairman of the Migratory Bird Advisory Board, stated that former friends had shunned him after publication of the article. It is unclear why Burnham waited so long to file the suit, but he had been thinking about it for at least a year. "I am bringing this not for personal vindication but because I can see no other way to clear the conservation atmosphere of the country," Burnham grandly wrote to his son on August 10, 1926. "I do not like this kind of case and would not go into it except for public spirited reasons." Perhaps, the gun-makers' withdrawal of their financial support from the AGPA earlier in the year contributed to the timing of the lawsuit. Hornaday countered that all his comments related to Burnham's public roles as a government advisor and president of a conservation organization. After initially offering to settle out of court, Hornaday reneged. Burnham assumed this was "probably as a result of financial and other support from 'zoophiles.'" In reality, it had everything to do with Hornaday's pugnacious nature and personal loathing of Burnham.[39]

Hornaday's defense, however, fell flat. The judge struck out large sections of his statement and made several adverse rulings in his court report. "A publication which imputes political corruption, or the use of political influence or privileges for pecuniary gain, is libelous, *per se*, even though the party against whom the charge is made is not a public officer or a candidate for office," Judge Heffernan wrote in response to Hornaday's claim that his statements did not libel Burnham. As for Hornaday's insistence that he had only laid the facts before the reader, the judge wrote, "Certainly the language is calculated to injure plaintiff's reputation and to degrade him in public estimation." Nor was Heffernan interested in litigating the case for bag limits in his court. Worse, Burnham held evidence that Hornaday had lied about not making the libelous statement, claiming instead that the *Times* inserted it without his knowledge. "Hornaday therefore swore falsely, probably having been told that he had put his foot in it," Burnham surmised. Facing obvious ignominy, Hornaday agreed to refrain from making intemperate remarks in public about John Burnham, a moderate, if difficult penalty. Yet, less than a year later, Burnham's lawyer had to remind Hornaday of their agreement after the director emeritus made yet more strong comments against his client in the *New York Times*.[40]

In December 1927, a horrific case of sciatica crippled Hornaday, a pain he described as "savage and fairly continuous." He spent the next six months bedridden, a terrible ordeal in itself for a man so normally active. Physical prostration coincided with gloom for the future of the nation's wildlife. "What is the use of any longer trying to protect any American game on a permanent basis?" he asked. "Here in the East, the combine against the game is, politically and financially, all powerful, and it's good night to the game." His physical condition captured the wider state of wildlife conservation. Just as experimental radium treatments cured Hornaday of his near paralysis, it would take novel forces to revive the wildlife conservation movement.[41]

9

FIGHTING TO THE END

IN JANUARY 1928, as William Temple Hornaday lay in his bed at the Anchorage, his home in Stamford, Connecticut, battling the "demon" sciatica, Senator Peter Norbeck, a South Dakota Republican associated with the progressive wing of the party, introduced another migratory bird refuge bill. Sick as he was, Hornaday remained a vigilant sentinel for wildlife, and he offered some unsolicited advice to the senator. "I have just finished a careful study of it," Hornaday wrote to Norbeck about his bill, "and I am convinced that it is not at all the good-conservation bill that you think it is." Hornaday's main gripe was the public-shooting-grounds provision. "The actual effect of the bill will be to put the federal government into the business of *founding and maintaining duck shooting resorts*," he wrote, "and generally promoting the *killing* of waterfowl." As an accommodation, Norbeck promised Hornaday to amend the bill to limit but not eliminate the shooting grounds. The number of sanctuaries without any shooting grounds "will run from sixty to ninety per cent," Norbeck opined.[1]

Carlos Avery, a close associate to John Burnham in the AGPA, the organization created by the gun makers, warned Norbeck against making any concessions to Hornaday, "because if you do there will be not much left of the bill." Norbeck himself had little interest in any of these provisions and considered them only irritating distractions from the main goal of securing bird refuges. "I am for the Sanctuary bill with or without shooting grounds," Norbeck pointedly commented to Hornaday's closest friend and ally, Edmund Seymour.[2]

Conservationists were not the only ones trying to shape the bill.

Norbeck's colleagues in the Senate demanded more amendments and modifications. In response, he reluctantly removed the feature of funding refuge acquisitions through hunting licenses in favor of funding them through general treasury appropriations. In two weeks' time, the bill Norbeck originally proposed had radically transformed into the version more fitting Hornaday's preferences, with no public shooting grounds and funding through general appropriations. The bill passed the upper chamber unanimously, an outcome that Hornaday described as "an AMAZING MIRACLE." Norbeck, however, was far from exultant. "I am frankly sick over the whole situation," he wrote to Edmund Seymour. "I have wasted so much effort on this matter and I look upon it as the biggest failure I have ever been connected with." Norbeck was not the only one who felt a keen sense of failure. The day after passage, John Burnham recognized his failure to achieve a public-shooting-grounds measure that would have allowed hunting on designated portions of the refuges by resigning as president of the AGPA, a bonus to Hornaday.[3]

Hornaday spent much of 1928 writing in support of the House passing a duplicate of the Norbeck bill. Despite the "miracle" in the Senate, his pessimism dictated that the sportsmen would attempt one last effort to restore the public-shooting-grounds and hunting-license provisions in the lower chamber and seal their victory in conference. "Any amendment now fastened upon that bill is the same as a ball and chain put upon a spent swimmer," he wrote in November. In December, he sent out copies of the first edition of *Plain Truth,* a wildlife conservation circular formatted as a newspaper that he published annually on his birthday. In it he likened the looseness with which his fellow citizens regarded Prohibition to an equally unserious attitude toward wildlife. Hornaday distributed copies of *Plain Truth* to magazines and legislators. Just in case anyone tried to ignore the message, he called the *New York Times,* and, as always, they gave him all the copy he wanted. On his seventy-fourth birthday, the *Times* ran "Game Refuge Bill Urged by Hornaday," an article for which he obviously was the only source. "For two months and more I have been working my hands and head off to boost your bill, arouse sleeping men and organizations, and head off opposition," Hornaday wrote to Norbeck the following month. As some insurance, he engaged a freelance journalist with whom he had worked in the past named Irving Brant to serve as his lobbyist in Washington.[4]

Hornaday's fears of amendment proved unfounded. Both the Audubon Society and the rump AGPA supported the House version of the Norbeck bill that passed 219 to 0 on February 9, 1929. Audubon's T. Gilbert Pearson, who had long supported the migratory bird bill with Burnham's public-shooting-grounds provision, to Hornaday's chagrin, came out wholeheartedly in favor of the Norbeck bill and organized a mass telegram and letter-writing campaign that resulted in more than two hundred thousand supportive missives on Capitol Hill. Some of these, however, might have emanated from an appeal by Hornaday's friend and ally Thornton Burgess, who had asked the audience of his *Radio Nature Program* on January 30 to write their representatives in support of the Norbeck bill. President Calvin Coolidge signed it into law on February 18. Within a decade of passage, the Norbeck Act created eighty-five refuges totaling 695,533 acres.[5]

Alone among the conservationists, Hornaday got all that he wanted out of the migratory bird refuge bill. More than anyone else, he had shaped the final product away from one that benefited sportsmen to one that granted equal access to wildlife to all. He was sure the hunters would gain from the refuges, but by removing the public shooting grounds from the legislation, Hornaday deprived them of their priority access to the waterfowl. And by blocking the license feature, he ensured that the hunters would not be able to use their power of the purse to determine the selection of the refuges. With funding from the general treasury the entire population had a stake in the waterfowl.

On September 12, seven months after passage of the bill, Will and Josie celebrated their golden wedding anniversary. "Each day I humbly thank the gods, for thee, my peerless wife, Who fifty years, through hopes and fears, Hath blessed my daily life," Hornaday wrote in a poem to his wife. Their daughter, Helen, did not plan an elaborate party, but wrote friends and family "to conspire in assembling a goodly number of letters from friends far and near on this, their Golden Wedding Anniversary." The Associated Press commemorated the event with a story, and the *New York Tribune* dispatched a reporter to interview the happy couple. Never one to hold his opinions to himself, Hornaday took the opportunity to tell the reporter that he disagreed with a notion gaining popularity that women were not good teachers for American boys because they were feminizing their charges. "I don't think the American man is going to be feminized to his injury," Hornaday opined. "If he selects the right wife, the more he listens to her advice the better off he is."⁶

In 1929, Willard Van Name, W. DeWitt Miller, and Davis Quinn wrote an incendiary pamphlet attacking the National Association of Audubon Societies. Crafted in the style of Hornaday's *Our Vanishing Wild Life* and forecasting imminent extinctions of specific species, *A Crisis in Conservation* attacked the NAAS, even though it did not mention it by name, for failing in its mission to protect birds. It charged the "rich sleepy private bird protection organizations" with misusing money, ignoring harmful changes in state wildlife laws, failing to lobby for protection of the bald eagle, and supporting the public-shooting-grounds provision through the early versions of the migratory bird refuge bills. By consorting with such vicious pro-hunting interests, the trio of authors argued, organizations like the National Association of Audubon Societies had betrayed their founding purposes and the desires of their large but mostly disenfranchised membership. It was a rare piece of work on wildlife conservation that impressed Hornaday. He considered it "a scorching pamphlet" and a "masterpiece of protest propaganda." *A Crisis* was merely the opening salvo in the most divisive struggle in the history of the Audubon Society, far more bitter and protracted than the 1911 imbroglio over the gun money. *A Crisis* also demonstrated that a younger generation found inspiration in Hornaday's stand against the public-shooting-grounds proposal in Burnham's version of the migratory bird refuge bill. Hornaday's calls to protect wildlife for purposes other than shooting it resonated with the younger generation, and they imitated his fiery, confrontational style, unbending moralism, and eagerness to challenge conservation organizations.⁷

The pamphlet inspired, among others, one Rosalie Edge, an activist who in time would make her own significant contributions to protecting American wildlife. Her interest in conservation mushroomed at the age of fifty-one after she read *A Crisis*. As an Audubon life member, she decided to attend the annual meeting in October 1929 at the American Museum of Natural History in New York to ask if the charges in the pamphlet were true. "My entrance made a stir, though no one knew me," Edge recalled. "That was the trouble; no stranger was expected." The Board of Directors, seated at a dais in the front, thought they could stifle any discussion of *A Crisis* at the start of the meeting with a general dismissal of the allegations. But Edge, the strong-willed daughter of a prominent New York family and a suffragette crusader, could not be deterred so easily. She rose and pointedly asked if the unnamed conservation organization that *A Crisis* claimed had done such a miserable job of protecting birds was, in fact, the Audubon Society. Edge felt like the skunk at the garden party as many in attendance scowled and stared at her. One member approached her and whispered to her that if the Audubon Society acknowledged the pamphlet, it "would surely call it to the attention of Dr. Hornaday," who would take "horrid advantage of the situation." It was a misplaced fear. Hornaday, who had more knowledge of what was going on in the public and private worlds of wildlife conservation than any of his contemporaries, had known about *A Crisis* days after its publication months before (and perhaps even before publication). After Edge's sharp questioning, Pearson huffily declared an end to the morning session, notified the attendees that the lady had drawn out the session with pointless questions, according to Edge's account, and hastily cancelled the afternoon film. "Two things were clear to me as a result of the meeting," Edge recollected: "the Directors were afraid of publicity, and they were afraid of Dr. Hornaday." Clearly, they were still smarting from the exposure of the gun manufacturers' subscription in 1911.[8]

Hornaday delighted in the stir that Edge created at the Audubon meeting. He dashed off a quick invitation to her to join him for lunch, praising her for possessing "the courage of a lion." Although more than twenty years divided them in age, the two forged a bond at their first meeting at his home in Stamford. Hornaday recounted in detail his own past dealings with Pearson and the Audubon Society, no doubt underscoring his many disappointments with their policy. Edge found her new mentor to be a "kindly, generous, humorous, charming, and learned old gentleman . . . filled with the fire of his enthusiasm, his burning desire to save wild life." Hornaday allowed her access to his private papers, including his scrapbook on the 1911 Audubon scandal over the gun money, as he had done for Irving Brant in 1927. Hornaday's voluminous collection of news clippings, articles, correspondence, memoranda, and scrapbooks was becoming an important archival resource for the challenger of the established order. "Dr. Hornaday was my mother's guide and mentor, teaching her the ins and outs of the problems, and incidentally introducing her to the important people involved, particularly in Washington," Peter Edge, Rosalie's son, commented.[9]

Within a month of the Audubon meeting, Rosalie Edge joined forces with Willard Van Name, one of the *Crisis* authors, to form the Emergency Conservation Committee (ECC), using her home address as the group's address. Hornaday declined an invitation to join the new organization, although he assisted it "generously with money and with counsel," according to Edge, and helped build the mailing list. The ECC allowed Van Name to evade a clause in his contract with the American Museum of Natural History that forbade him from publishing pieces unless they first passed muster with a review panel at the American Museum of Natural History. "They can prevent me from signing them," Van Name told Edge, "but they cannot prevent my writing them." Irving Brant was asked to join the ECC after Edge read his article in *Forest & Stream* supporting *A Crisis* and condemning the Audubon leadership. Thus, the ECC consisted of three members, two of whom Hornaday had personally influenced. Each would share for decades the ideology he preached, and Brant continued to serve as the Permanent Wildlife Protection Fund's lobbyist in Washington.[10]

Around the same time that Edge attended the Audubon annual meeting, Senator Charles McNary, an Oregon Republican, introduced Hornaday's bag-limit-reduction bill into the upper chamber. Three weeks later, Representative Gilbert Haugen, an Iowa Republican, did likewise in the House. Hornaday was feeling uncharacteristically optimistic. "Just between ourselves," he wrote to Brant, "I think that the passage of the McNary-Haugen bill is now *assured!*" But the Eastern Establishment, including the Biological Survey, resisted. "Such a limit should be considered and finally established by people who have experience with wildlife and its destruction," the old conservationist George Bird Grinnell reasoned, "and the number of Congressmen who have had such experience is of course extremely small." Grinnell may have had a point, but Hornaday introduced the bill not because he thought it was the best method of administration, but because the scientific advisors of the Migratory Bird Advisory Board recommended a bag-limit reduction and a politician serving as secretary of agriculture, Anthony Hyde, had refused to implement the changes.[11]

On the last day of 1929, Secretary of Agriculture Anthony Hyde, the choice of newly inaugurated President Herbert Hoover, on his own accord reduced the bag limit. Harry McGuire of *Outdoor Life*, who supported Hornaday on the migratory bird bill but was bothered by the hostility aroused within the wildlife-protection camp over the past decade, considered it a victory. "The great principle he has been fighting for has been recognized," McGuire said. But the view from Stamford, Connecticut, was very different. Hornaday believed that Hyde had acted only to neutralize the McNary-Haugen legislation and allow the Biological Survey to save face in a difficult situation. He demanded that his congressional sponsors decisively move forward with the legislation. "We must savagely press the passage of the bill to have the bag limit fight settled for ten years not merely one just now," he wired Brant. With the economic outlook growing dimmer by the day and unemployment roles swelling, however, Congress had scant time to focus on ducks. Brant informed Hornaday at the end of February that the McNary-Haugen bill was about dead. Hornaday's chance, if it ever really existed at all, of enshrining a bag limit by statute lay in ashes. "There is nothing new in Washington!" he declared in disgust to Edmund Seymour in March 1930.[12]

In October, Hornaday was once again drawn into a controversy at the Audubon Society. After Irving Brant wrote his first pamphlet for the ECC entitled *Compromised Conservation: Can the Audubon Society Explain?*, a scathing condemnation of the society's policy positions and leadership timed to coincide with the annual meeting, Audubon's leadership attacked Brant in a pamphlet of their own. But Brant achieved the effect that Edge had wanted. "I thought it would stimulate attendance and discussion," Edge wrote in her unpublished "Autobiography." "It did." Although Hornaday remained close to the ECC, the group shielded him from this particular pamphlet. "The tendency of the Audubon directors is to blame Dr. Hornaday for all criticisms of the Association," Brant later wrote, and thus he wanted to maintain as much distance as possible. Reform had to be separated from personalities, and this was one possible way to do it. Nonetheless, Hornaday greatly enjoyed Brant's work. "Your Compromised Conservation manifesto is a great piece of constructive conservation. . . . I hope it marks the end of Benedict Arnoldism in that organization," he telegrammed.[13]

Edge asked Hornaday to join her for the Audubon annual meeting on October 28, 1930. He mailed in five dollars for a membership and thereby the right to attend. As soon as the general business portion of the meeting concluded, Hornaday rose from his chair and read his seven-part resolution calling for the Audubon Society to endorse a series of measures, including the McNary-Haugen bag-limit bill. It was not a surprise: he had written Audubon's treasurer, Robert Cushman Murphy, the week before to inform the Board of Directors of both his intent to speak and the topics. Never comfortable speaking before public groups, and still uneasy about it, Hornaday nonetheless felt this speech was the best one he had ever given. "The 150 members sat dumb—afraid to vote for our resolution and also afraid to vote it down!" Hornaday himself recalled in *Thirty Years War for Wild Life*. The directors referred the resolution and the questions raised by Brant in "Compromised Conservation" to a special three-man committee consisting exclusively of men friendly to Pearson. For the larger membership, Audubon's *Bird-Lore* trivialized the episode, writing of Hornaday's resolutions: "The Secretary, William P. Wharton, called attention to the fact that the majority of these referred to lines of endeavors on which the Association had long been active." Murphy denied Brant's allegations that Pearson received a commission from memberships. Either Murphy told a bald-faced lie or he was totally out of his depth in dealing with the financial affairs of the association. Ever since 1922, Pearson had collected a commission on memberships, exactly as Brant charged.[14]

The events of the day greatly troubled Edge. She described her feelings after the meeting as "depressed and tired," and Hornaday's serenity surprised her. Unlike Brant and Edge, he held out little hope that the Audubon Society could be reformed, and he possessed little interest in doing so. For twenty years, Pearson had consistently stood with the killers of game, men like John Burnham and Carlos Avery, and, by extension, the merchants of death themselves. No, the outcome of the meeting did not bother him at all. It only confirmed what he had known for twenty years. He was done with the Audubon Society and did not involve himself again in Edge's crusade to change the group from within, although he gave whatever help she asked for. She continued in the ensuing years to question the Audubon budget and its policies in preserves that it owned, and waged a proxy ballot battle against Pearson. Each year she came back with new arguments and allegations. In January 1933, the investment banker John Baker succeeded in nullifying Pearson's power. Pearson remained on as window dressing before exiting the scene entirely in September 1934. Baker reoriented the Audubon Society in a more progressive direction from the perspective of the ECC. Edge's persistence had paid off. She had proven Hornaday wrong: Audubon could be reformed.[15]

The Audubon grandees never fully understood the connection Edge had forged with Hornaday and clumsily attempted to sever it. The most personal effort took place in October 1931, when George Pratt, who described Hornaday as "extremely unreasonable and unfair in his attacks on certain conservationists," sent Edge extracts from *Two Years in the Jungle,* hoping to shock her out of Hornaday's orbit. Edge refused to fall for such a cheap shot. "I cannot agree with you," she responded, "that any man may not attain wisdom after he is thirty." Edge was probably a little insulted as well. She was a determined individual with a clear sense of right and wrong who saw many legitimate problems at the Audubon Society headquarters. She did not need Hornaday or anyone else to provide moral guidance on such issues. Hornaday thanked Edge for defending him from another of Pratt's attacks. "In closing," he wrote to her, "I repeat, and urge you to remember the fact that I am not a repentant sinner in regard to my previous career as a killer and preserver of wild animals, but I am positively the most defiant devil that ever came to town. I am ready and anxious to match records for my whole 76 years with any sportsman who wishes to back his record against mine for square dealing with wild animals." Although Hornaday declined to attend any Audubon meetings after the 1930 event, his continuing financial and moral support proved valuable to Edge. "It is always darkest before the dawn," Edge wrote to Hornaday in receipt of a check in 1932, "and *you* are the Dawn!"[16]

Edge accepted Hornaday more uncritically than Brant, who understood his employer and mentor's shortcomings more clearly. "You have to know how the doctor's mind works," Brant wrote to Edge in 1931. "Other people simply swim into his ken. He will take a certain issue at a certain date, and the people of that issue and date divide into heroes and villains. . . . Hornaday is entirely incapable of surveying a field without reference to himself. That shortcoming, of course, is part of his strength, which grows out of his self-centeredness." Edge seems to have never seen this side of Hornaday. Instead, according to her son Peter Edge, the author of *Our Vanishing Wild Life* "gave great stability to her work and [she] did not have to prove herself." This connection earned Edge the moniker of "sentimentalist" from the establishment, a label she shared with Hornaday and a sign that the two spoke of wildlife as more than an economic resource.[17]

Brant continued to attack the Biological Survey for its connection to sportsmen in ECC pamphlets that resonated with his mentor's logic and ideology. Hornaday influenced the content of Brant's writings, encouraging his protégé to remove attacks on the Boone and Crockett Club in deference to an aged and ill Madison Grant, among other such suggestions. "I used most of the quotations you sent Mrs. Edge," Brant wrote Hornaday in July 1931, "and put in Pearson's record on the automatic shotgun." Brant's writings resembled Hornaday's to such a degree that Pearson told Frank Chapman that they "are identical to the type of Hornaday's writings and most things he says are in Hornaday's publications." This was no surprise. The upcoming journalist had studied his conservation history in Hornaday's private archives and, of course, worked for the Permanent Wildlife Protection Fund. Brant decried "shotgun conservationists" and demanded an end to the practice of baiting animals when hunting. Clearly, this recalled all of Hornaday's efforts over the last thirty years to limit shotgun use, only to be scoffed at by sportsmen all along the way.[18]

In 1931, Charles Scribner's Sons published Hornaday's *Thirty Years War for Wild Life: Gains and Losses in the Thankless Task*. The title says it all. Daniel Carter Beard, founder of the Boy Scouts, accurately described this latest and final original addition to the Hornaday canon as "enlightening, interesting, and belligerent." Despite the conservation achievements of the past, it seemed by Hornaday's account as if wildlife was more endangered than ever before. More hunters took to the field each year, while the means of conservation lagged far behind the constantly evolving tools of death. The old market hunters had been put down, but new economic interests had replaced them. Camping outfitters like L. L. Bean had taken the place of game-meat dealers like August Silz, but the killing went on. In all, Hornaday calculated hunting in America to be a $300 million a year industry. As in the past, he emphasized the role technology played in fueling the killing. "The outstanding fact in the progressive extinction of American game birds is that 90 per cent. of it has been accomplished by the gun-and-automobile combination, and only 10 per cent. of it is due to other causes."[19]

Hornaday dedicated his book to "the Congress of the United

States as a small token of appreciation of its generous services to wild life during the decade from 1920 to 1930, in new legislation to provide game sanctuaries, and to reduce excessive killing privileges." The dedication captured a fundamental precept of Hornaday's beliefs. As a lobbyist and policy wonk, he always felt real reform could only come from the American people themselves. Though he could single out a few politicians for condemnation, he never blamed politicians as a whole for failing to pass better laws. Like the rebirth and redemption he had heard about at the Adventist revivals on his father's farm, true reform, Hornaday believed, could only come from the hearts of each individual. In the case of baiting, for example, no law should be required to stop Americans from engaging in an immoral form of hunting. But his view of hunters had not improved. "It is my estimate that 15 per cent. of the *sportsmen* are humane and reasonable conservationists," he wrote.[20]

President Herbert Hoover, a fellow Iowan, pleased Hornaday in 1931 with two favorable and long-awaited decisions on issues close to the doctor's heart. In June, Hoover declared a moratorium on all intergovernmental Allied war debts from the First World War. Hornaday thought it morally unjust for the United States to demand repayment on the loans. Instead, he felt the United States should give the Allied countries the money for fighting the war while his own country evaded its responsibility to protect civilization against German militarism. In 1928, he had attended a conference in New York with the purpose of eliminating the debts, but the effort failed to gain any traction. President Calvin Coolidge, whom Hornaday admired in other ways, had failed on this particular issue. It had taken a global economic collapse to persuade Hoover to consider a debt moratorium as a tool to "assist in the reestablishment of confidence, thus forwarding political peace and economic stability in the world." Hornaday considered the moratorium a sound idea and one that would win over the president's growing number of enemies.[21]

President Hoover followed his debt moratorium with an equally significant decision on wildlife conservation. On July 3, 1931, the Agriculture Department issued a grim forecast for the fall, one that, if accurate, would have dire consequences for America's ducks, geese, and wildfowl. "Concern for the safety of the birds is increased by announcements from the Weather Bureau that the extreme deficiency of moisture that was experienced in 1930 followed similar conditions in the Northwest dating back to 1922," the department's statement read. A *New York Times* reporter predicted that "this year's hatch of waterfowl may be the smallest on record." The Agriculture Department's pronouncement hit Hornaday like a thunderbolt, and he implored President Hoover to greatly curtail the hunting season as the only solution to the crisis. "In view of the awful slaughter by drought during 1929 and 1930, it is needless for me to point out to you what this third one means," he wrote Hoover. The deadly effect of habitat loss reinforced his belief that overhunting was the primary driver of wildlife extinction. After all, there was little the government could do in the short term to increase habitat. He offered no long-term solution to drought because he feared imminent extinction for many species. Hunting restrictions, however, could mitigate the crisis more immediately. A reduction in the shooting, according to Hornaday's calculations, would save "probably 1,000,000 of the drought-stricken and weakened ducks and geese" and earn the president adulation from "50,000,000 American and Canadian people." Whether out of political affinity or shared heritage in the Hawk Eye State, Hornaday wrote more freely to Hoover than to any president since his friend Theodore Roosevelt had occupied the White House. Hoover's acting secretary of agriculture, R. W. Dunlap, echoed Hornaday's concern to the president. "The Department shares the alarm indicated by Dr. Hornaday in respect to migratory waterfowl," Dunlap stated. On August 25, 1931, President Hoover issued a proclamation shortening the waterfowl hunting season to a mere month. Hornaday praised Hoover's "ready sympathy and courageous initiative" in the December 1 edition of *Plain Truth*.[22]

Yet, with the economy worsening by the day, and with a presidential election at the end of the year, 1932 was not an auspicious time for bold wildlife-protection measures. Hornaday remained as convinced as ever that the nexus of overhunting and commercialism was exterminating the game. "There are at least 50 per cent. too many guns in commission," he wrote in *Plain Truth*. "They are 50 per cent. too deadly in execution. . . . Clustering closely around the power-houses of the gun and ammunition industries, like remoras on sharks, are the allied Commercial Interests."[23]

Hornaday issued *A Call to Action* before hearings on wildlife chaired by Senator Frederick Walcott in the spring of 1932. Hornaday assumed Walcott's intent was to undo Hoover's proclamation and undermine enhanced hunting regulations enacted by the secretary of agriculture. To counter this possibility, Hornaday called for the extension of the shortened waterfowl-hunting season to the following year. He also demanded an end to baiting and a cartridge tax to add money to the federal coffers for refuge acquisition and employment of wardens. He preferred a cartridge tax to a license because the more a hunter shot, the more he paid. "Among other things," he wrote, "it is vitally important to the existence of migratory game that the federal government should *vigorously participate in its defense!*" Hornaday was unable to get his cartridge tax through the Walcott Committee, but at least the sportsmen failed to roll back any regulations. He would gladly take this policy wash over a defeat.[24]

Despite the recent favorable actions of the federal government, Hornaday remained critical of the connection between sportsmen and policy makers. John Burnham, that lackey of the hunters and gun makers, may have passed from the scene, but the publisher George Knapp, who founded Ducks Unlimited, had taken his place as wildlife's public enemy number one. "Knapp is a dangerous man for the game of North America," Hornaday charged. After briefly suspending the Migratory Advisory Board, Secretary of Agriculture Arthur Hyde reconstituted the panel and ordered that all meetings be held in secret. Hornaday could not have been intentionally baited any better. On behalf of the Permanent Wildlife Protection Fund, Irving Brant spent some time trying to crack the closed proceedings to see what ugly secrets were hidden behind those sealed doors. After two years of research, Brant had enough material to write an ECC pamphlet ridiculing the advisory council as the "Reactionary Advisory Board" because it "was more representative of the point of view of these reactionary eastern sportsmen than the old one." Much to Hornaday's disappointment, regulations for the 1932–33 hunting season expanded its duration to eight weeks instead of reducing it. A marginal cut in the bag limit from fifteen to twelve ducks and a closed season for Brant geese on the East Coast were of little consequence to Hornaday, who craved radical action.[25]

In November 1932, amid failing banks and rising unemployment, the nation went to the polls to elect a president. Hornaday cast his ballot with the minority. "Roosevelt is a good man, and a gentleman," Hornaday wrote of FDR, "but just now I don't want to see any swapping of horses in the middle of a turbulent stream." Aside from his desire to maintain continuity of leadership during a crisis, he preferred

Hoover's pro-tariff and gold standard policies. Although they appear not to have ever met, Hornaday had many reasons to sympathize with the embattled Hoover. Both were sons of Iowa and self-styled orphans who had traveled the world. And Hornaday, too, considered himself under siege for advocating unpopular, if correct, reforms. In mid-October, Hornaday thought the president was improving in the polls. "I think Hoover is now gaining," he optimistically wrote. But Roosevelt trounced Hoover in a landslide. "I am too tired to think about our wild life protection campaign," Hornaday wrote Edmund Seymour a week after the presidential election, "but I do know that the general situation is 90 per cent hopeless."[26]

Hornaday began channeling increasing amounts of his time into writing a full autobiography, "Eighty Fascinating Years." At various times he tentatively titled it "Bringing Wildlife to the Millions," "A Life of Joyous Work," "The Lure of Fascinating Work," and "Unbeaten Trails through Life." Hornaday claimed in "An Introduction to Myself" that Henry Fairfield Osborn had inspired him to write an account of his life. When Hornaday was recovering from a second eye surgery in 1922, Osborn had solicited a promise from his recuperating director, "As soon as your recovery is complete, I want you to write that autobiography." Hornaday had created a partial manuscript entitled "Evolution of a Zoologist" a few years later. He took up the project in earnest after his retirement, but *Thirty Years War for Wild Life*, a mixture of conservation career highlights and demands for new reforms, muscled out a more genuine autobiography. In contrast, "Eighty Fascinating Years" contained little on his conservation career, focusing instead on his life before 1896 and his nonconservation activities afterward. Hornaday drew amply on his already published material; large chunks came from his books and articles.[27]

"Eighty Fascinating Years" was never published. Hornaday tried to sell it to Charles Scribner's Sons in the spring of 1935, but they either rejected it or suggested too many editorial changes, because he went to another publishing firm later in the year. After his death, George and Helen Fielding attempted without success to find a publisher. It is hard to pin down exactly why no printing house saw fit to publish the memoir of one of the nation's foremost conservationists and naturalists. The fact that most of it had been published already in various formats might have been the biggest deterrent.[28]

There were two chapters in Hornaday's draft commenting on contemporary society and politics marked "not for publication." Chapter 20, entitled "World Mistakes," captures the seismic change in Hornaday's mind-set around World War I. "Thirty years ago I was a sincere optimist on the impulses and good faith of humanity, and the moral fiber and intelligence of civilized man," he wrote. "Today, I think that speaking generally, Civilized Man is an unmitigated ass." He worried over Nazi designs on Europe. "By his persistent appeals to their greed and egotism, Hitler has made them [Germans] believe that another big war would be a great thing for Germany,—and them!" Hornaday wrote. "Today the rabid war insanity of Germany seems destined to bear terrible fruit,—and in the near future." Hornaday's fears, of course, proved to be justified, as the Second World War claimed the lives of 60 million people, including one of his beloved grandsons.[29]

Hornaday followed up his caustic depiction of the global scene with an equally grim description of American culture. This chapter, "A Slump in American Morale," was an edited version of the convocation he gave to the students at Ames in 1927. Not surprisingly, he railed against the vices of the Jazz Age. "Young men of America, for God's sake snap out of your sport cars, ye [sic] speakeasies and your roadhouses, and brace up for the reform of the nation of 50 per cent slackers and criminals," he exhorted. "Get away from those blaring and yawping radios, and the printed trash of the 'popular' newsstands, and do some serious reading and thinking." Despite coming off as a cranky old man suspicious of new ways of thinking, Hornaday had a much deeper and poignant point. He earnestly believed that civilization had taken a disastrous turn in 1914 with the start of World War I, and that the young people of the 1920s and 1930s possessed not the slightest inkling of what life had been like before the cataclysm. "In the spring of 1914, both socially and economically this world as a whole was at the highest point it ever has attained," he wrote, "or ever is likely to attain in the future." Hornaday did not necessarily blame the kids for exercising bad morals; in his mind, they were influenced by a corrupt alliance of immigrant-based organized crime and the American Civil Liberties Union (ACLU) who protected them. "The serpents from Europe that we have taken into our bosoms during the fifty years prior to 1914 are now fighting us and pouring poison into our veins to an extent so serious that presently it may become fatal," he added. He blamed the ACLU and their fellow travelers for being more concerned with the rights of criminals than with the common good. Hornaday had already gone on record as favoring a concealed-carry gun law so that law-abiding citizens could arm themselves, and for scrubbing the Constitution clean of the Fourth and Fifth Amendments, which he believed only protected the criminal class.[30]

If he was worried about American youth in general, Hornaday never felt so nervous about his own grandchildren. He was proud of his granddaughter Lorraine's independence and accomplishments. In 1928, he gave her a vacation at a dude ranch as a birthday present. She recounted her experiences in a book, *French Heels to Spurs*. By the time he wrote "Eighty Fascinating Years," she was a successful model with photographs in several popular publications. Hornaday's grandsons Temple and Dodge both entered Princeton University. Their grandfather generously paid their tuition and sent them an allowance for living expenses out of his pension from the New York Zoological Society.[31]

As for his old nemeses, hunters, Hornaday testified before an Agriculture Department special hearing on baiting in August 1933. He walked into hostile territory. Many at the hearings found a *New York Times* editorial he had written earlier in the month to be offensive. "It seems to a majority of the advisory board that masquerades as the Federal official watchmen and advisors of Secretary Wallace will not stand for any stoppage of the three killing privileges [baiting of game, use of live decoys, and sale of game 'on the wing' by large sporting clubs] of which the defenders of game bitterly complain," Hornaday acerbically wrote. Primed to retaliate against Hornaday's editorial, one New Jersey game official jumped up when Hornaday spoke and demanded that "Dr. Hornaday be made to stick to the truth." This and other interruptions might have broken his concentration, as his testimony veered wildly off topic several times. "Very well," Chairman Thomas Beck motioned, "in view of the doctor's health and age, I move that we extend him every courtesy and permit him to ramble on." It may have been the worst performance of Hornaday's life.[32]

If the baiting hearings failed to move the needle on wildlife protection, a presidential committee consisting of Thomas Beck, Aldo Leopold, and Jay "Ding" Darling revived interest in migratory bird refuges and obtaining adequate funding for them. Once again, the idea of using hunting licenses as a funding source emerged. Hunters would be required to purchase a one-dollar duck stamp to hunt migratory birds, with the proceeds earmarked for the acquisition of refuges under the

Norbeck Act. Hornaday considered it a dangerous idea. "The boosting of the Duck Stamp Tax Bill will admirably serve as a blind from which to defend the shotguns," he opined to Edmund Seymour in December 1933. "Our conclusion on the Duck Stamp Tax bill is that it is visionary and lacking in large merit for the protection of more game BY BREEDING," he commented in the seventh edition of *Plain Truth*. But this feeble and mostly rhetorical opposition could not prevent the Migratory Bird Stamp Act from becoming law in early 1934. "When the Roosevelt administration came in," Hornaday wrote later, "in less than one year it stole the plan and purpose of the Norbeck bill, tagged it with the name of Bawling Beck of Connecticut, and put it forth with a great flourish of trumpets as the Roosevelt Administration's child."[33]

After signing the Duck Stamp law, President Franklin Roosevelt appointed Jay "Ding" Darling chief of the Biological Survey. It was an odd choice. A professional political cartoonist and conservative Republican, he possessed neither scientific credentials nor political affinity for the administration. Hornaday knew Darling well. The cartoonist provided drawings for Hornaday's published works, and the two fellow Iowans corresponded. Friendship, however, did not correlate exactly with influence, and Hornaday felt Darling had intentionally and needlessly constructed a barrier between them. "Ding has been very careful to give me no advance information on anything, and to ask my advice on nothing," Hornaday wrote in July 1934. A month later, he described Darling as "inexperienced and credulous" and "silly about the wisdom" of the secretive Migratory Bird Advisory Board. In other words, he had become a tool of the game hogs. Darling defended himself. "The great body of sportsmen themselves are the worst problem we have had to contend with," he wrote to Edmund Seymour. "I came down here to Washington with the idea that I was going to stand up for what was right and tell everybody to go to hell who opposed sane methods of conservation. I still feel the same way. No one has tied my hands."[34]

Hornaday turned eighty on December 1, 1934, and his birthday was a news event. Some old friends from the zoo ventured to Stamford to pay their respects, including W. Reid Blair, Lee Crandall, Raymond Ditmars, and H. R. Mitchell. One newspaper noted that Hornaday still "carries on his fight for the preservation of wild life with as much vigor and enthusiasm today at the age of 80 as he did half of a century ago." Hornaday used the publicity to focus attention on his most recent issue of *Plain Truth* entitled "The Wild Ducks' Waterloo," where he made several dire assertions regarding the fate of wildlife, including predictions of imminent extinction.[35]

In February 1935, Darling officially recommended to the secretary of agriculture and the president that conditions "indicate the necessity for drastic restrictions." He recommended a short season instead of a closed season, reduced bag limits, and a maximum of three shells in each shotgun. President Roosevelt quickly assented.[36]

Hornaday should have been delighted at the implementation of a policy he had been demanding for three decades. "The long-looked-for reduction in the killing capacity of magazine shotguns has at last been realized," the Bureau of Biological Survey press release opened. His deep-rooted pessimism on the future of wildlife in America got the better of him. "They came at half past the eleventh hour, when migratory game began to look like a total loss," he wrote months later in *Plain Truth*. Nonetheless, he considered the changes a triumph of those wildlife protectionists who had pressured the government, especially Irving Brant and the ECC. "Mrs. Edge and Hornaday congratulated me on the effects of my pamphlets," Brant wrote in his memoirs. "Nobody, it seems, thought to congratulate Franklin Roosevelt, who took time out from a herculean economic-recovery task to grasp and perform a job in conservation by action that antagonized most of his wealthiest friends and enemies."[37]

In February 1936, two weeks after making his final public appearance on behalf of wildlife conservation at the American Wildlife Conference, Hornaday entered the hospital for tests to determine the nature of an illness that afflicted him. It had been a difficult month. "Fate has been giving us some punches on the chin recently," he wrote his grandsons Dodge and Temple on the eve of entering the hospital, "these two hospitalizations, the wrecked car, and now a big plumbing job in our bathroom!" Despite these difficulties, he still paid Dodge's and Temple's allowances and tuition at Princeton, and took care to write their checks in advance so that they would have spending money on hand. The doctors who treated Hornaday drew no conclusions from the battery of tests and could offer no real prognosis. "My doctor pretends to think I will survive it all and get well," he wrote in March 1936, "but I begin to doubt it. Well, in any event, this totally unfair and accursed ailment can not rob me of the 81 years of good health that I have had, and I can say to the last minute that as to 'Life' I have Lived, and had a mighty good run for my money! And so, to hell with the old needlessly-crippled legs!" Despite his attempt to be philosophical, Hornaday realized full well that he was entering the final days of his life. He did not expect to be long of the earth.[38]

In September, Hornaday made arrangements with W. Reid Blair, his successor at the New York Zoological Park, for the transference of his wildlife conservation papers, including his correspondence and scrapbooks, to the New York Zoological Society. "These volumes have been carefully kept down to the limits of useful history," Hornaday wrote, "for the benefit of historians who twenty years from now will be scratching around to find the answer to the question, 'why did the American people permit all their game to be exterminated?'" He had already signed documents to transfer the Permanent Wildlife Protection Fund over to the New York Zoological Society on his death.[39]

The following month he became ill with the sickness that would eventually take his life after five months in bed. But Hornaday put on a good show for the media and others who visited him on his birthday in December, and the local paper, the *Stamford Advocate*, carried an account of the festivities. He made yet one more appeal for restraint on the part of hunters, asking "all gentlemen who are sportsmen, and all sportsmen who are gentlemen, not to shoot waterfowl next year, so that the year 1940 will not witness the practical extinction of United States waterfowl and waterfowl hunting in the United States and southern Canada." The usual delegation of dignitaries from the New York Zoological Society had trucked up to Stamford one more time to pay their respects to their old boss.[40]

Hornaday died on March 6, 1937. He passed quietly in the evening at his home, the Anchorage. It was front-page news for the *Stamford Advocate*. Obituaries all praised his wildlife conservation work, but their coverage of other parts of his career varied widely. "Had it not been for the foresight and energy of Dr. Hornaday, and others who joined him in his campaign, the game birds of the United States would be nearly if not quite extinct," wrote the *Indianapolis Star*. The *New York Herald Tribune* noted that Hornaday was "an ardent anti-Bolshevist" and had "attacked [the] domestic cat" for its killing of birds. The *Indianapolis Star* emphasized his collecting trips abroad.[41]

The funeral service followed on Tuesday, March 9, at the First Presbyterian Church in Stamford. William Temple Hornaday was laid to rest at Putnam Cemetery in Greenwich. A singer from Stamford sang

"Home on the Range," Hornaday's favorite song, and recited "Trees" by Joyce Kilmer. New York Zoological Park keepers served as pallbearers, and the Boy Scouts provided an honor guard, including one scout who had earned two Hornaday Awards. There were eighteen honorary pallbearers, including Daniel Carter Beard, W. Reid Blair, Fairfield Osborn, Willard Van Name, and Frederic Walcott. Madison Grant was also present. No fewer than two hundred Bronx Zoo employees attended the service, including Lee Crandall and Raymond Ditmars. Hornaday's wife, Josephine, would follow her husband on January 16, 1939.[42]

EPILOGUE

ON APRIL 18, 1912, George Orville Shields pulled his copy of William Temple Hornaday's *Two Years in the Jungle* off the shelf for a little evening reading. After thirty minutes, he took out his pen to write the book's author, his old friend. "I have kept you busy explaining and apologizing for me," Shields wrote, "but we may both console ourselves with the assurance that 100 years hence people will understand both of us better than they do now, and will be sorry that our warnings were not heeded more generally than they are."[1]

Actually, within the next twenty-five years, a younger generation would view Hornaday as a needlessly polarizing presence and outmoded thinker. Members of this generation of conservationists noticed that Hornaday's predictions of imminent extinctions, like William Miller's prophesies of Christ's Second Coming in 1843 and 1844, all too often failed to materialize. In giving Hornaday the very last speaking slot in President Franklin Roosevelt's five-day wildlife-protection conference in 1936, the younger generation acknowledged his irrelevance to them.

In his lecture "The Last Call for Game Salvage," Hornaday did not deviate from his message of the past fifty years. Now a wizened eighty-two-year-old, he unrepentantly reiterated the central message of his conservation career. "Let no man or men talk to me about other causes than shooting by sportsmen for game decrease or extinction," he told the remaining delegates. Hornaday was clearly out of place amid the cooperative mood and shared desire of the delegates to move past the strife-ridden 1920s and reconstruct a broad, viable wildlife conservation coalition. The conference endorsed both the Duck Stamp Act, to which Hornaday never reconciled himself, and the Darling regulatory reforms of 1935, which he favored.[2]

The delegates reacted positively to the Roosevelt administration's push for unification of the wildlife-protection movement. They formed the General National Wildlife Federation, which was renamed the National Wildlife Federation (NWF) the following year. Although dominated by sportsmen's groups, it was a far cry from the elitist Boone and Crockett Club and the gun maker–sponsored American Game Protection Association (AGPA) that had so infuriated Hornaday. The NWF included a significant contingent of nonhunters within its ranks, who bridged the gap between the utilitarian and preservationist wings of the movement.[3]

The effort at reconciliation at the conference required compromises that the older generation of conservationists had previously been unable to make as they fought bitterly over key features of the Norbeck Act. But compromise would allow both sides to gain something they wanted. While Hornaday would have been gratified had he lived to see the bald eagle finally receive protection in 1940, he would have been horrified that public shooting grounds were permitted in migratory bird refuges. In an odd twist, the Norbeck Act (one of Hornaday's greatest triumphs) had been completely undone. A hunter-financed duck stamp had replaced payments from the general treasury to acquire bird refuges. The addition of public shooting grounds to the refuges in 1949 and their expansion in 1966 marked a thorough repudiation of Hornaday's stand throughout the 1920s. Compromises among wildlife-protection advocates were needed to make these changes.

But not all of Hornaday's pet issues met the same fate as the migratory bird refuges. His proposed regulations on baiting animals, use of live decoys, and the number of shells a hunter could carry in his shotgun were implemented in 1935 and have remained on the books for more than seventy years. If Congress rejected his sanctuary plan in 1916 and 1917, states enacted versions of their own, thanks to the lobbying efforts of local organizations.

Buffalo refuges have fared well since the establishment of the Wichita refuge in 1905. Although the current population of about four hundred thousand buffalo in the United States is dwarfed by the tens of millions of buffalo that existed when Columbus landed, the current number vastly exceeds the numbers that roamed in the wild when Hornaday undertook the "last scientific buffalo hunt" in 1886.

The seals prospered after the restoration of harvesting in the 1920s, and reached 2 million in 1950, before a new enemy, pollution, triggered another population crisis. Commercial harvesting ceased entirely in the 1980s. Protected by the Marine Mammal Protection Act of 1972, the seal population numbered 750,000 in 2007, more than twenty times higher than when Henry Wood Elliott begged Hornaday to do something to save them a century earlier.[4]

If Hornaday's conservation efforts benefited wildlife, his ideology left an imprint on the younger generation, even if they rejected his confrontational and combative style. Aldo Leopold echoed many of Hornaday's criticisms of the consumer economy and sportsmen in an essay in *A Sand County Almanac* entitled "Wildlife in American Culture," which first appeared as an article in 1943 in the *Journal of Wildlife Management*.

"The conquest of nature by machines has led to much unnecessary destruction of resources," Leopold wrote. "Our tools improve faster than we do." In another passage, he criticized the new economic interests that profited from the killing of wildlife, again much in the manner of Hornaday: "The traffic in gadgets adds up to astronomical sums." Leopold even attacked sportsmen for failing to live up to their own ideal of voluntary self-restraint, as Hornaday had done, and asked, "Where is the go-light idea, the one-bullet tradition?" And, like Hornaday, Leopold connected hunting to the larger consumer economy and the industry that profited from the killing. "The sportsman has no leaders to tell him what is wrong," Leopold bemoaned. "The sporting press no longer represents sport, it has turned billboard of the gadgeteer." Leopold's more modest style obscures the connection with Hornaday, whose strident Progressive Era moralism, pointed attacks (he would have named those sportsmen, their organizations, and the magazines), and exact quantification of costs resounded more clearly in the writings of Irving Brant, Thornton Burgess, Rosalie Edge, and Willard Van Name, and in the cartoons of Jay "Ding" Darling.[5]

Hornaday also influenced future generations in another way. Through his zoo, numerous writings, and conservation campaigns, he expanded the scope of animals that deserved protection beyond economically valuable "game" to include the more inclusive concept of "wildlife." While he would perhaps have been puzzled by late-twentieth- and early-twenty-first-century efforts to propagate wolves (a species he described as belonging to the "criminal" class) and return them to the wild, he charted a course that successive generations have taken to enlarge the meaning of "wildlife." Moreover, in valuing preservation over immediate economic benefits, Hornaday served as a harbinger of the later environmental movement. The course that he forged and others followed came to fruition with the landmark Endangered Species Act of 1973, which has been used to save owls and snails, among many other species never mentioned in the discourse of the Progressive Era.

Hornaday may have been wrong when he asserted that all wildlife in America would be extinct by 1950, but the threat of extinction still looms over many species. In June 2012, Lonesome George, the last survivor of the Pinta tortoise of the Galapagos Islands, a species that inspired Charles Darwin to formulate his theory of evolution, died. Like the death of Martha the passenger pigeon ninety-eight years before, Lonesome George's passing extinguished something that can never be rekindled. As Hornaday well knew, extinction is permanent, which is the reason the current generation needs to be on guard against it.

NOTES

Abbreviations

HAW	Henry Augustus Ward
HAWP	Henry Augustus Ward Papers
JCH	Josephine Chamberlain Hornaday
NYT	New York Times
ODP	Office of the Director (William Hornaday) Papers
OPMGP	Office of the President (Madison Grant) Papers
OSMGP	Office of the Secretary (Madison Grant) Papers
WTH	William Temple Hornaday
WTHP-GT	William T. Hornaday Collection, Guilford Township Historical Collection
WTHP-LC	William Temple Hornaday Papers, Library of Congress
WTHP-WCS	William Temple Hornaday Papers, Wildlife Conservation Society

INTRODUCTION

1. There is a voluminous literature describing the rapid decline of wildlife in America. I recommend Busch, *The War Against the Seals;* Isenberg, *The Destruction of the Bison;* Kimball and Kimball, *The Market Hunter;* Matthiessen, *Wildlife in America;* Roe, *The North American Buffalo;* Schorger, *The Passenger Pigeon;* Tober, *Who Owns the Wildlife?;* Trefethen, *An American Crusade;* and Wilcove, *The Condor's Shadow.*

2. On conservation constituencies, see Fox, *John Muir and His Legacy;* Graham, *The Audubon Ark;* Hays, *Conservation and the Gospel of Efficiency;* and Reiger, *American Sportsmen and the Origins of Conservation.*

3. Miller, *Gifford Pinchot and the Making of Modern Environmentalism*, 139–44; Nash, *Wilderness and the American Mind*, 161–81; Worster, *A Passion for Nature*, 403–32.

4. WTH to Charles Vorhies, 1 February 1913, Outgoing Correspondence, vol. 4, WTHP-WCS.

1. IOWA FARM BOY

1. Q. Hornaday, *The Hornadays Root and Branch*, 3–4; WTH, "Eighty Fascinating Years," box 17, WTHP-LC. Recent research by Patsy Ann Combs Hornaday indicates that it is possible that Hornaday is a Dutch name and that they received land in Northern Ireland as a reward for service by Charles II, not Cromwell. See http://members.cox.net/annelwood/Finding%20Hornadays.htm (courtesy of Steven Hornaday).

2. Fischer, *Albion's Seed*, 613–787; Q. Hornaday, *The Hornadays*, 5–8.

3. Q. Hornaday, *The Hornadays*, 64–65, 255–56.

4. WTH, "Eighty Fascinating Years," box 17, WTHP-LC; Q. Hornaday, *The Hornadays*, 261. Margaret died in 1856.

5. WTH, "Eighty Fascinating Years," box 17, WTHP-LC; ibid., box 15.

6. Ibid., box 17, WTHP-LC; *History of Hendricks County, Indiana*, 753.

7. WTH, "Eighty Fascinating Years," box 17, WTHP-LC.

8. Q. Hornaday, *The Hornadays*, 261 (Silas was a son of William's first wife, Orpha Hadley); both of Martha's quotes are from Martha Varner Hornaday to her parents, 17 March 1859, box 1, Michael H. Miller Collection; WTH, "Eighty Fascinating Years," box 15, WTHP-LC.

9. Martha Varner Hornaday to her parents, 17 March 1859, box 1, Michael H. Miller Collection; Martha Varner Hornaday to her mother, dated 26 August, no year, Minos Miller Papers.

10. WTH to W. Jamieson, 14 September 1912, Personal Letter Book, vol. 11, box 80, WTHP-LC.

11. Martha Varner Hornaday to her parents, 11 July 1863, box 1, Michael H. Miller Collection.

12. Minos Miller to Martha Varner Hornaday, 18 January 1863, Minos Miller Papers; Minos Miller to Martha Varner Hornaday, 17 August 1866, ibid.

13. Martha Varner Hornaday to David Miller, 22 February 1864, Michael H. Miller Collection.

14. WTH to David Miller, 4 May 1864, box 1, Michael H. Miller Collection; Martha Varner Hornaday to David Miller, 5 May 1864, ibid.

15. WTH quoted in Dolph, "Bringing Wildlife to the Millions," 6; WTH to David Miller, 19 February 1864, box 1, Michael H. Miller Collection; WTH to Samuel Shortridge, 2 February 1926, Outgoing Correspondence, reel 15, WTHP-WCS.

16. J. B. Grinnell, *Events of Forty Years*, 113; Gaustad, ed., *The Rise of Adventism*; Land, ed., *Adventism in America*; Morgan, *Adventism and the American Republic*.

17. WTH, *Free Rum on the Congo and What It Is Doing There*; Morgan, *Adventism and the American Republic*, 35–37, 64–71; WTH to Chester Jackson, 13 November 1876, Chester Jackson Papers.

18. Bryan and Bryan, *The Memoirs of William Jennings Bryan*, 1:36.

19. WTH, "Evolution of a Zoologist," manuscript, ca. 1926, pp. 5, 8, 7, WTHP-WCS. He did not repeat these stories in "Eighty Fascinating Years."

20. Martha Varner Hornaday to David Miller and Minos Miller, 11 February 1860, Minos Miller Papers.

21. Martha Varner Hornaday to William Hornaday, Monday morning 1866, box 43, WTHP-LC; WTH, "Eighty Fascinating Years," box 15, p. 12, WTHP-LC.

22. WTH, "Eighty Fascinating Years," chap. 2:1, 2, WTHP-WCS. This version is a completed, although disorderly, manuscript. Citations to "Eighty Fascinating Years" that include the chapter and page number refer to the version at WTHP-WCS.

23. WTH, "Eighty Fascinating Years," box 16, p. 13, WTHP-LC.

24. WTH to Louis Pammel, 29 April 1926, Louis H. Pammel Papers, University Archives, Iowa State University Library, Ames; for Bessey, see Overfield, *Science with Practice*.

25. WTH to Charles Edwin Bessey, 10 November 1906, Charles Edwin Bessey Papers.

26. WTH, "Eighty Fascinating Years," chap. 2:3-4.

27. WTH with Holland, *Taxidermy and Zoological Collecting*, 63; WTH, "Eighty Fascinating Years," chap. 2:2; ibid., chap. 2:6; WTH with Holland, *Taxidermy and Zoological Collecting*, 63.

28. WTH with Holland, *Taxidermy and Zoological Collecting*, 64.

29. WTH, "Eighty Fascinating Years," chap. 2:9; WTH to Theodore Roosevelt, 10 October 1916, Theodore Roosevelt Papers.

30. WTH, "Eighty Fascinating Years," chap. 2:9; Barrow, "The Specimen Dealer," 493-534; WTH to HAW, 11 April 1873, HAWP; WTH to HAW, 28 April 1873, ibid.

31. WTH to HAW, 16 August 1873, HAWP; WTH to HAW, 18 October 1873, ibid.

32. WTH, "Eighty Fascinating Years," chap. 3:1; Lucas, *Fifty Years of Museum Work*, 10; WTH to Charles Edwin Bessey, 14 December 1873, box 11, WTHP-LC.

33. WTH, "Eighty Fascinating Years," chap. 3:2; Ward, *Henry A. Ward*, 177; WTH, "A Great Museum Builder," 31-32.

34. WTH, "Eighty Fascinating Years," chap. 4:1 (numbered p. 52).

35. Jenkins, *The Naturalists*, 99-107.

36. WTH, "Eighty Fascinating Years," chap. 4:1, 2 (numbered pages 52-53).

37. WTH, "Eighty Fascinating Years," chap. 4:7 (numbered p. 58).

38. WTH, *Wild Animal Interviews*, 173-74.

39. WTH to HAW, 8 January 1875, HAWP; WTH to HAW, 14 January 1875, ibid.; WTH to HAW, 5 February 1875, ibid.; 28 February 1875, ibid.; WTH to HAW, 15 March 1875, ibid.

40. WTH, "The Crocodile in Florida"; Nathaniel H. Bishop, "The Florida Crocodile," *Forest & Stream* 13 (1 January 1880): 947; "An American True Crocodile," *Forest & Stream* 4 (22 April 1877): 167; Cope, *Crocodilians, Lizards, and Snakes of North America*, 175.

41. WTH to HAW, 5 February 1875, HAWP; WTH, "Eighty Fascinating Years," chap. 3:63.

42. Cutright, *The Great Naturalists Explore South America;* "Memoranda of Agreement made this day between Prof. Ward and Mr. Hornaday," 25 January

1876, box 88, WTHP-LC; HAW and Chester Jackson, "Memo of Agreement," 20 January 1876, box 14, ibid.

43. Dolph, "Bringing Wildlife to the Millions," 83–86.

44. Quoted ibid., 90–91.

45. Chester Jackson, "Journal," pp. 7, 8, 10, Chester Jackson Papers; WTH to HAW, 6 March 1876, HAWP; WTH, "Eighty Fascinating Years," box 15, 64–69, WTHP-LC; WTH, *A Wild Animal Round-Up*, 140–41.

46. WTH to HAW, 1 April 1876, HAWP; C. Jackson, "Journal," Chester Jackson Papers, 16; WTH to HAW, 25 May 1876, HAWP; WTH to Theodore Roosevelt, 18 May 1914, Theodore Roosevelt Papers.

47. C. Jackson, "Journal," Chester Jackson Papers, 24, 35; WTH to HAW, 24 June 1876, HAWP; WTH to HAW, 4 July 1876, ibid.

48. WTH, "Eighty Fascinating Years," chap. 3:64–70, 72, 73–75. Jackson recorded a less dramatic account of the rescue from quicksand in his journals. He did not mention the presence of Indians or indicate that Hornaday was in mortal danger (Chester Jackson, "Journal," Chester Jackson Papers, 18).

2. COLLECTING NATURALIST AND HUNTER

1. WTH, *Two Years in the Jungle*, 1.

2. WTH to JCH, 20 March 1900, box 1, WTHP-LC; WTH to Chester Jackson, 15 December 1878, Chester Jackson Papers.

3. "Memoranda of an Agreement between Henry A. Ward and W. T. Hornaday," October 1876, HAWP.

4. Rangarajan, *Fencing the Forest;* MacKenzie, *Empire of Nature*.

5. WTH, *Two Years in the Jungle*, 3; WTH to Chester Jackson, 13 November 1876, Chester Jackson Papers; WTH to Josephine Chamberlain, 19 October 1877, box 1, WTHP-LC.

6. WTH, *Two Years in the Jungle*, 4; WTH to Chester Jackson, 13 November 1876, Chester Jackson Papers.

7. WTH, *Two Years in the Jungle*, 6.

8. Ibid.

9. Ibid., 13, 14, 15, 17.

10. Ibid., 19. Hornaday referred to Ross's rank at the time he wrote *Two Years in the Jungle*.

11. WTH to HAW, 19 February 1877, HAWP. So that the chapter text remains consistent with quotations, I employ the place-names used by Hornaday. The current place-names are Mumbai (Bombay), Kolkata (Calcutta), Sri Lanka (Ceylon), Varanasi (Benares), and Yamuna (Jumna).

12. WTH, *Two Years in the Jungle*, 22; WTH to HAW, 19 February 1877, HAWP.

13. WTH, *Two Years in the Jungle*, 183; WTH to HAW, 17 March 1878, HAWP; WTH, *Two Years in the Jungle*, v, 65; James, *Raj*, 304.

14. WTH, *Two Years in the Jungle*, 29.

15. Ibid., 49–50.

16. WTH to Josephine Chamberlain, 25 June 1877, box 1, WTHP-LC; WTH, *Two Years in the Jungle*, 79.

17. WTH, *Two Years, in the Jungle*, 84, 86.
18. Ibid., 96; WTH to HAW, 10 June 1877, HAWP; WTH with Holland, *Taxidermy and Zoological Collecting*, 8.
19. WTH to HAW, 10 June 1877, HAWP.
20. WTH, *Two Years in the Jungle*, 111; WTH to HAW, 10 June 1877, HAWP; WTH to JCH, 25 June 1877, box 1, WTHP-LC; Dolph, "Bringing Wildlife to the Millions," 173.
21. WTH to HAW, 25 June 1877, HAWP; WTH, *Two Years in the Jungle*, 123; WTH to HAW, 14 July 1877, HAWP; WTH to HAW, 4 August 1877, ibid.; WTH to Chester Jackson, 17 October 1877, Chester Jackson Papers.
22. WTH to HAW, 26 July 1877, HAWP.
23. WTH to HAW, 15 September 1877, HAWP; WTH, *Tales from Nature's Wonderlands*, 185; WTH to HAW, 17 October 1877, HAWP.
24. WTH to HAW, 22 September 1877, HAWP; WTH, *Two Years in the Jungle*, 164, 165. Hornaday did not mention to Jackson that he had poached the elephant, but he used the code "peculiar" to convey this (WTH to Chester Jackson, 17 October 1877, Chester Jackson Papers).
25. WTH to Chester Jackson, 17 October 1877, Chester Jackson Papers; WTH to HAW, 20 November 1877, HAWP.
26. Ibid.
27. WTH to HAW, 3 December 1877, HAWP; Dolph, "Bringing Wildlife to the Millions," 198; WTH, *Two Years in the Jungle*, 208–17.
28. WTH to Frederic Lucas, 12 December 1877, quoted in Dolph, "Bringing Wildlife to the Millions, 201; WTH to HAW, 20 December 1877, HAWP.
29. WTH, *Two Years in the Jungle*, 290; WTH to HAW, 4 February 1878, HAWP.
30. WTH to HAW, 14 January 1878, HAWP; WTH, *Two Years in the Jungle*, 241; WTH to Chester Jackson, 10 February 1878, Chester Jackson Papers.
31. WTH to HAW, 4 February 1877, HAWP; WTH to Chester Jackson, 10 February 1878, Chester Jackson Papers; WTH to HAW, 17 March 1878, HAWP.
32. WTH with Holland, *Taxidermy and Zoological Collecting*, 71; WTH, *Two Years in the Jungle*, 283; WTH to HAW, 2 May 1878, HAWP.
33. WTH to HAW, 31 May 1878, HAWP; WTH to HAW, 4 June 1878, ibid.
34. WTH to Chester Jackson, 27 May 1878, Chester Jackson Papers.
35. WTH to HAW, 4 June 1878, HAWP; WTH to HAW, 11 July 1878, ibid.
36. WTH to HAW, 18 July 1878, HAWP; WTH to HAW, 26 July 1878, ibid; WTH to Chester Jackson, 2 February 1879, Chester Jackson Papers.
37. WTH to HAW, 26 July 1878, HAWP.
38. WTH to HAW, 5 August 1878, HAWP; WTH to HAW, 21 August 1878, ibid.
39. WTH to HAW, 30 August 1878, HAWP.
40. WTH, *Two Years in the Jungle*, 354.
41. WTH to HAW, 21 October 1878, HAWP.
42. WTH to HAW, 22 December 1878, HAWP; WTH to Chester Jackson, 2 February 1879, Chester Jackson Papers.
43. Carnegie, *Round the World*, 159; WTH, *Two Years in the Jungle*, 297.
44. WTH, "Eighty Fascinating Years," chap. 15:4.

45. WTH to Chester Jackson, 10 June 1879, Chester Jackson Papers; WTH to HAW, 21 May 1879, HAWP; WTH to HAW, 2 December 1879, ibid.
46. WTH to Chester Jackson, 10 June 1879, Chester Jackson Papers; Carnegie, *Round the World*, 159.

3. STUFFED AND LIVING ANIMALS

1. WTH, "Eighty Fascinating Years," chap. 8:39; WTH to HAW, 18 July 1878, HAWP.
2. WTH, "Eighty Fascinating Years," chap. 8:38–39; WTH, "Fighting among Wild Animals."
3. Irmscher, *The Poetics of Natural History*, 75; Sellers, *Charles Willson Peale*.
4. WTH, "On the Species of Bornean Orangs," 438–55; WTH, "Eighty Fascinating Years," chap. 8:39, chap. 9:2.
5. WTH to JCH, 31 August 1902, box 1, WTHP-LC; WTH to JCH, 27 May 1902, ibid.
6. WTH to Mary Hornaday, 7 October 1881, box 2, WTHP-LC; WTH to David Miller, 17 April 1884, WTHP-GT; WTH to Helen Ross Hornaday, 23 May 1897, box 2, WTHP-LC.
7. WTH, "Eighty Fascinating Years," chap. 8:40; Bledstein, *The Culture of Professionalism*.
8. Society of American Taxidermists, *Third Annual Report*, 7.
9. WTH with Holland, *Taxidermy and Zoological Collecting*, 232; Lucas, "The Mounting of Mungo," 337.
10. WTH to Charles Edwin Bessey, 10 September 1907, Charles Edwin Bessey Papers; WTH to HAW, 30 July 1882, HAWP; WTH to HAW, 15 October 1883, ibid.; Madden, *The Authentic Animal*, 129–32.
11. Society of American Taxidermists, *Third Annual Report*, 31, 33; WTH to HAW, 20 July 1882, HAWP; WTH to HAW, 10 September 1882, ibid.; WTH to HAW, 4 October 1882, ibid.; WTH to HAW, 2 December 1882, ibid.; WTH to HAW, 28 December 1882, ibid.
12. WTH, "Eighty Fascinating Years," chap. 15:5.
13. WTH, "Masterpieces in American Taxidermy," 6–7.
14. WTH, "Eighty Fascinating Years," chap. 3:2.
15. Ibid., chap. 8:46; WTH to HAW, 4 May 1882, HAWP.
16. WTH to David Miller, 17 April 1884, WTHP-GT; WTH to HAW, 4 May 1882, HAWP; Hilkey, *Character Is Capital;* Dolph, "Bringing Wildlife to the Millions," 369–70; WTH to Albert Bigelow Paine, undated, ca. 1903, Letters to Albert Bigelow Paine, Huntington Library, San Marino, Calif.
17. Madden, *The Authentic Animal*, 132; Board of Regents, *Annual Report, 1884*, 659–70.
18. Charles Scribner's Sons to WTH, 19 September 1885, box 43, WTHP-LC; "Hornaday's Travels in Borneo," *Science* 6 (27 November 1885): 472–73; S.C.C., "Two Years in the Jungle," *Forest & Stream* 26 (4 February 1886): 22.
19. WTH, *A Wild Animal Round-Up*, 5; WTH, "The Extermination of the American Bison," 529; WTH, *A Wild Animal Round-Up*, 5, 7.

20. WTH, *A Wild Animal Round-Up*, 6.

21. Roosevelt, *Hunting Trips of a Ranchman*, 244; WTH, *A Wild Animal Round-Up*, 9–10, 7–8.

22. WTH, *A Wild Animal Round-Up*, 11–15; "The National Museum Buffalo," *Forest & Stream* 28 (23 April 1887): 106.

23. Peterson, "Buffalo Hunting in Montana in 1886," 6; WTH, "The Extermination of the American Bison, 413, 534; WTH, *A Wild Animal Round-Up*, 15–18.

24. Peterson, "Buffalo Hunting in Montana in 1886," 7; WTH, *A Wild Animal Round-Up*, 27; Punke, *Last Stand*, 137; Dolph, "Bringing Wildlife to the Millions," 539; WTH, *A Wild Animal Round-Up*, 18–28.

25. WTH, *A Wild Animal Round-Up*, 42; WTH, "Extermination of the American Bison," 502; Gard, *The Great Buffalo Hunt*, 102–3.

26. WTH, *A Wild Animal Round-Up*, 50–51; WTH, "The Extermination of the American Bison," 532.

27. WTH to Spencer F. Baird, 21 December 1886, http://siarchives.si.edu/history/exhibits/documents/21Dec1886Letter.htm; Board of Regents, *Annual Report, 1887*, 60; WTH, *A Wild Animal Round-Up*, 43–53.

28. "A Big Buffalo Hunt," *Decatur Illinois Saturday Herald*, 1 January 1887, www.ancestry.com; Peterson, "Buffalo Hunting in Montana in 1886," 11.

29. WTH, "The Extermination of the American Bison," 547; WTH, "Eighty Fascinating Years," chap. 16:11.

30. WTH, "Eighty Fascinating Years," chap. 20:1; WTH, "What Price Dead Ducks, 33.

31. Roe, *The North American Buffalo*, 7, 530–42; Flores, "Bison Ecology and Bison Diplomacy," 465–85; WTH, "The Extermination of the American Bison," 499; Dolph, "Bringing Wildlife to the Millions," 539. See also Trefethen, *An American Crusade*, 22; Isenberg, *The Destruction of the Bison*, 141–43.

32. Grinnell and Sheldon, eds., *Hunting and Conservation*, 219, 32–33.

33. WTH, "Extermination of the American Bison," 464–65, 435, 525, 391. Two other ideas did not stick as well: "the phenomenal stupidity" of the animal and the preference for shooting cows.

34. WTH, "Eighty Fascinating Years," chap. 9:1; WTH to George Brown Goode, 25 June 1887, box 43, WTHP-LC; WTH, "Eighty Fascinating Years," chap. 9:2.

35. Board of Regents, *Annual Report, 1888*, 214, 213–22.

36. WTH to George Brown Goode, 18 January 1888, box 43, WTHP-LC.

37. WTH, "Eighty Fascinating Years," chap. 9:3; Dolph, "Bringing Wildlife to the Millions," 593–94.

38. WTH, "Eighty Fascinating Years," chap. 9:7.

39. *Annual Report, 1889*, 28; WTH, "Eighty Fascinating Years," chap. 9:8.

40. WTH, "Eighty Fascinating Years," chap. 9:1, 10; *Congressional Record*, 1 April 1890, 2893; WTH, "Eighty Fascinating Years," chap. 9:14.

41. Horowitz, "The National Zoological Park," 418; Baker, "National Zoological Park," 454; WTH, "Eighty Fascinating Years," chap. 9:11–12, 11; Samuel P. Langley to WTH, 6 May 1890, box 43, WTHP-LC; Samuel P. Langley to WTH, 10 May 1890, ibid.

42. WTH, "Eighty Fascinating Years," chap. 9:14; Samuel P. Langley to WTH,

17 May 1890, box 43, WTHP-LC; Dolph, "Bringing Wildlife to the Millions," 640–41.

43. WTH, "Eighty Fascinating Years," chap. 9:15; WTH to Samuel P. Langley, 16 May 1890, box 43, WTHP-LC.

44. WTH, "Eighty Fascinating Years," chap. 9:16.

45. G. Bailey, *Illustrated Buffalo*, 88.

46. WTH, "How to Observe Quadrupeds," 415.

47. WTH, "Eighty Fascinating Years," chap. 8:44; WTH to HAW, 30 July 1882, HAWP; "Taxidermy and Collecting," *Forest & Stream* 37 (23 July 1891): 3; *Science* 18 (21 August 1891): 110–11.

48. WTH quoted in *Illustrated Buffalo*, 88; Allen Varner to David Miller, 17 July 1892, box 1, Michael H. Miller Collection; Dolph, "Bringing Wildlife to the Millions," 651.

49. *Book News* 14 (January 1896): 247; WTH, *The Man Who Became a Savage*, 5; WTH, journal, 1 February 1896, box 110, WTHP-LC.

4. DIRECTOR OF THE BRONX ZOO

1. WTH, journal entries, 6 January 1896, box 110, WTHP-LC; Henry Fairfield Osborn to WTH, 6 January 1896, box 44, ibid.

2. WTH, journal entries, 8 January, and 31 January 1896, box 110, WTHP-LC.

3. Henry Fairfield Osborn to WTH, 1 April 1896, box 44, WTHP-LC; WTH, journal entry, 3 April 1896, box 110, ibid.; WTH, journal entry, 5 April 1896, box 110, ibid.; WTH to JCH, 8 October 1896, box 1, ibid.

4. Deforest Grant, "History of the New York Zoological Society," ca. 1923, box 1, OPMGP.

5. Quoted in Brooks, *Speaking for Nature*, 127; WTH, "Eighty Fascinating Years," chap. 11:3.

6. "The New York Zoological Park," *Forest & Stream* 50 (1 January 1898): 1; Horowitz, "Animal and Man in the New York Zoological Park," 441; Roosevelt, *An Autobiography*, 324; Theodore Roosevelt to Deforest Grant, 4 December 1894 in Madison Grant, "Origins of the New York Zoological Society," January 1928, Zoological History Project, box 1, OPMGP, 1; Theodore Roosevelt to Madison Grant, 10 October 1894, *Letters of Theodore Roosevelt*, 1:401.

7. WTH, "The London Zoological Society and its Gardens," 44; WTH, "Report upon a Tour of Inspection of the Zoological Gardens of Europe," 35.

8. "The New Bronx Park Zoo," *NYT*, 2 September 1896.

9. WTH, European Journals, vol. 4, box 88, WTHP-LC; "Plans for the Proposed Zoological Park in New York," *Science* 4 (6 November 1896): 681–82; WTH, "A Report upon a Tour," 41.

10. "Report of the Executive Committee to the Board of Managers," *Annual Report of the New York Zoological Society* 1 (15 March 1897): 23–25; WTH, "Eighty Fascinating Years," chap. 15:5.

11. WTH, Journal, 1 May 1896, box 110, WTHP-LC; WTH, *A Wild Animal Round-Up*, 355, italics in original; WTH, "Report on Character and Availability of South Bronx Park," 26–34.

12. WTH, "The London Zoological Gardens," 65; WTH, "Report upon a Tour of Inspection of the Zoological Gardens of Europe," 38–39; "For a Zoological Park," *NYT*, 17 December 1896; Madison Grant, "History," 1909, Zoological History Project, box 2, OPMGP, 6.

13. WTH, "Report of the Director of the Zoological Park" (1899), 39, 40–41; Horowitz, "Animal and Man in the New York Zoological Park," 433.

14. WTH, "Report of the Director of the New York Zoological Park to the Board of Managers," 47–71; Welker, *Natural Man*, 10; WTH, "Twelve Years' Perspective of the Zoological Park," 109; WTH, *A Wild Animal Round-Up*, 180–82.

15. Merkel, "The New York Idea of a Zoological Park"; Crandall, *A Zooman's Notebook*, 3; Stott, "An American Idea of a Zoological Park," 5.

16. "New York Zoological Gardens," *NYT*, 24 September 1899; "New York 'Zoo' Formally Opens," *New York Herald*, 9 November 1899; "New York 'Zoo' Opened," *New York Tribune*, 9 November 1899; "Zoological Park Opened," *NYT*, 9 November 1899; Osborn, "Progress of the New York Zoological Park."

17. Bridges, *Gathering of Animals*, 127–28; Horowitz, "Animal and Man in the New York Zoological Park," 431.

18. WTH, "Report of the Director of the Zoological Park" (1904), 50.

19. WTH, *The Minds and Manners of Wild Animals*, 53, 75; J. Turner, *Reckoning with the Beast*, 74; WTH to JCH, 27 May 1902, box 1, WTHP-LC.

20. WTH to Henry Fairfield Osborn, 13 March 1909, box 30, OSMGP; Hancocks, *A Different Nature*, 101–3.

21. WTH to Charles Edwin Bessey, 15 December 1897, Charles Edwin Bessey Papers; WTH, "The Destruction of Our Birds and Mammals," 78.

22. WTH, "The Destruction of Our Birds and Mammals," 101; Palmer, *Legislation for the Protection of Birds Other Than Game Birds*, 41; *Auk* 15 (July 1898): 280–81; *NYT*, 17 and 24 April 1898; Cameron, *The Bureau of Biological Survey*, 24–42.

23. On the connection of race and wildlife conservation, see Dunlap, *Saving America's Wildlife*, 12.

24. WTH to Charles Davenport, 21 June 1898, Charles Davenport Papers, American Philosophical Society, Philadelphia; WTH, *Thirty Years War for Wild Life*, 153; WTH, "Eighty Fascinating Years," chap. 20:9.

25. Spiro, *Defending the Master Race*, 37; WTH to Henry Fairfield Osborn, 29 January 1920, Outgoing Correspondence, reel 9, WTHP-WCS.

26. WTH, "Rules and Regulations of the Zoological Park," 5 January 1900, ODP; WTH to JCH, undated, 1899, box 1, WTHP-LC. He later wrote to say he thought Ray would work out after all (see WTH to JCH, 24 May 1899, ibid.).

27. WTH, "Warning Against Gossiping," 5 November 1915, ODP; "Hirshfield Renews Hornaday Attack," *NYT*, 1 April 1920; WTH, "The London Zoological Society and its Garden," 67; Bridges, *Gathering of Animals*, 363–64.

28. WTH to the Employees of the New York Zoological Park, 30 January 1900, ODP; WTH, "Rules and Regulations of the Zoological Park," 5 January 1900, ibid.

29. WTH, "The Right Way to Teach Zoology," 260; Armitage, *The Nature Study Movement*.

30. WTH, The *American Natural History*, 28; WTH to Theodore Roosevelt,

29 May 1907, Theodore Roosevelt Papers; Lutts, *The Nature Fakers;* WTH, "The Right Way to Teach Zoology."

31. WTH to Samuel Marvin, 7 November 1903, Samuel Marvin Papers, Alderman Library, University of Virginia, Charlottesville; WTH to Samuel Marvin, 14 March 1903, ibid.; Reiger, *American Sportsmen and the Origins of Conservation,* 217; R. A. Franks to WTH, 8 March 1902, box 2, WTHP-WCS.

32. *Bulletin of the American Geographic Society* 42 (1910): 623–24; "A New Natural History," *Forest & Stream* 64 (11 February 1905): 112.

33. *American Naturalist* 39 (June 1905): 412, 413; WTH, The *American Natural History,* vi.

34. Walter K. Fischer, *Condor* 7 (September–October 1905): 45; Mary Cooney to WTH, 30 July 1910, box 2, WTHP-WCS; WTH, "Eighty Fascinating Years," chap. 16:4.

35. WTH's royalty reports are in box 33, WTHP-LC.

36. Spiro, *Defending the Master Race,* 63; WTH to John Franklin Lacey, 24 March 1906, John F. Lacey Papers, State Historical Society of Iowa; WTH, "The Wichita National Bison Preserve," 556.

37. Roosevelt, *An Autobiography,* 436.

38. Ibid., 408–36; Richardson, *A Compilation of the Messages and Papers of the Presidents,* 16:6911; Williams, *The Forest Service,* 140.

39. Theodore S. Palmer to Madison Grant, 21 March 1906, box 1, OSMGP; Theodore Roosevelt to John F. Lacey, 21 March 1906, Theodore Roosevelt Papers.

40. Loring, "The Wichita Buffalo Range," 181–200.

41. WTH to Madison Grant, 17 July 1906, box 1, OSMGP; James B. Adams to WTH, 15 August 1906, box 49, WTHP-LC; W. Swift to WTH, 15 August 1906, ibid.

42. "A New National Buffalo Herd," *Science* 26 (25 October 1907): 563.

43. John F. Lacey to Madison Grant, 25 July 1906, box 1, OMGSP; WTH to John F. Lacey, 15 September, 1906, Lacey Papers; "15 Buffalo to Go Back to the Range," *NYT,* 6 October 1907; *Year Book of the Department of Agriculture,* 584; Elwin R. Sanborn, "An Object Lesson in Bison Preservation," 990.

44. "Bison Preserves," *NYT,* 3 November 1907; Williams, *The Forest Service,* 140.

5. CAMPFIRES AND CONSERVATION

1. WTH to Chester Jackson, 24 August 1905, box 66, WTHP-LC; WTH to John M. Phillips, 24 July 1905, ibid.; Harper, "John McFarlane Phillips," 461–62; WTH, *Camp-Fires in the Canadian Rockies,* 1, 13, 106.

2. WTH, *Tales from Nature's Wonderlands,* 166; WTH to JCH, 30 September 1905, box 1, WTHP-LC.

3. Calvin Hornaday to WTH, 11 July 1906, box 49, WTHP-LC; WTH to T. E. Donne, 21 January 1920, Outgoing Correspondence, reel 9, WTHP-WCS.

4. Bradford and Blume, *Ota Benga;* Verner, "The Story of Ota Benga," 1377; R. Adams, *Sideshow U.S.A,* 35.

5. Spiro, *Defending the Master Race,* 45; Verner quoted in Bradford and Blume, *Ota Benga,* 173.

6. George Bird Grinnell to WTH, 27 March 1896; reel 4, George Bird Grinnell Papers; Bradford and Verner, *Ota Benga*, 174, 177; *NYT*, 9 September 1906; Ham and Ware, *Darwin's Plantation*, 19–20.

7. Washington, *Medical Apartheid*, 77–78; *NYT*, 13 September 1906; Henry Fairfield Osborn quoted in Bradford and Blume, *Ota Benga*, 189.

8. Bradford and Verner, *Ota Benga*, 174, 187, 189; "Suicide of Ota Benga, the African Pygmy," *New York Zoological Society Bulletin* 19 (May 1916): 1356.

9. WTH, "An African Pygmy"; "Suicide of Ota Benga," 1356; Bradford and Blume, *Ota Benga*, 191–232; Madden, *The Authentic Animal*, 131; Washington, *Medical Apartheid*, 77–78; Spiro, *Defending the Master Race*, 47–49.

10. WTH, *Camp-Fires in the Canadian Rockies*, vi, 145.

11. Shields, "Hunting Caribou with a Camera," 305; WTH, *Camp-Fires in the Canadian Rockies*, 189–98.

12. WTH, *Camp-Fires in the Canadian Rockies*, vii; A. Bryan Williams to WTH, 11 January 1907, Scrapbook, vol. 1:9, WTHP-WCS; WTH, *Thirty Years War for Wild Life*, 207; Runte, *National Parks*; undated comment in Scrapbook, vol. 1:10.

13. A. Bryan Williams to WTH, 25 February 1907, Scrapbook, vol. 1:14; Colpitts, *Game in the Garden*, 134–36; undated handwritten comment, p. 15, ibid.; WTH, *Thirty Years War for Wild Life*, 209; WTH, *Camp-Fires in the Canadian Rockies*, 89; A. Bryan Williams to S. Dennis (vice president of the Canadian Pacific Railroad), 4 March 1907, Scrapbook, vol. 1:16; WTH and John M. Phillips to Board of Trade, 5 May 1907, p. 42, ibid.

14. WTH, *Thirty Years War*, 207, 207–9; A. B. Mackenzie to WTH, 20 March 1907, Scrapbook, vol. 1:30; WTH to Premier McBride, 12 December 1906, p. 22, ibid.; Runte, *National Parks*, 48–64; Colpitts, *Game in the Garden*, 136.

15. Colpitts, *Game in the Garden*, 136–138; Herchemer to John M. Phillips, 29 February 1908, Scrapbook, vol. 1:73; "Fine Game Reserve in East Kootenay," *Victoria Daily Colonist*, 14 October 1908.

16. WTH, *Camp-Fires on Desert and Lava*, vii; WTH to David Miller, 25 August 1907, WTHP-GT.

17. WTH to David Miller, 25 August 1907, WTHP-GT.

18. Osborn, *The American Museum of Natural History*, 68; WTH to Madison Grant, 27 April 1907, box 1, OSMGP; SMGP; Madison Grant to WTH, 9 May 1907, box 3, ibid.; Madison Grant to Henry Fairfield Osborn, 30 April 1907, ibid.; Madison Grant to Henry Fairfield Osborn, 13 May 1907, box 1, ibid. It is unclear how the matter of the antlers was resolved.

19. WTH, *Camp-Fires on Desert and Lava*; Bill Boyles, "Adventures in the Pinnacate," *Journal of Arizona History* (Summer 1987), copy in box 1, WTHP-WCS.

20. WTH, *Camp-Fires on Desert and Lava*, 92–95; WTH to Theodore Roosevelt, 1 January 1907, Theodore Roosevelt Papers (correct date 1908; year misstated by author); Theodore Roosevelt to WTH, 10 December 1907, Theodore Roosevelt Papers; "Hornaday Visits Arboreal Desert," *New York Saturday Review of Books*, 28 November 1908, 712.

21. WTH, *Camp-Fires on Desert and Lava*, 171, 15, 38, 29.

22. Ibid., ix, 15; WTH, "Director of Zoo Makes Protest," *NYT*, 29 May 1908; WTH, "Notice to Keepers and Janitors," 10 October 1907, ODP.

23. Trefethen, *An American Crusade*, 138; Baynes, "The Fight to Save the Buffalo," 133; Pearson, *Adventures in Bird Protection*, 133.

24. "Another Reservation for Buffalo," *Washington Star*, 4 March 1908, Scrapbook, col. 2, WTHP-WCS; WTH, "Eighty Fascinating Years," chap. 15:11, 12; Sterling, *Last of the Naturalists*, 273; WTH to Charles Davenport, 16 April 1908, Davenport Papers; WTH to Boies Penrose, 30 March 1908, Scrapbook, vol. 2; Charles C. Scott to WTH, 16 May 1908, ibid.

25. WTH to Morton Elrod, 6 April 1908, Personal Letter Book, vol. 5, box 78, WTHP-LC; Charles Aubrey to George Bird Grinnell, 8 January 1908, reel 56, George Bird Grinnell Papers; Charles Aubrey to George Bird Grinnell, 1 June 1908, ibid.; Morton Elrod to WTH, 11 January 1908, box 2, WTHP-WCS; WTH to Michael Pablo, 21 May 1908, Personal Letter Book, vol. 5, box 78, WTHP-LC; Isenberg, "The Returns of the Bison," 188. "For a Buffalo Preserve," *New York Evening Post*, 6 March 1908, Scrapbook, vol. 2:1; Charles Aubrey to George Bird Grinnell, 26 July 1908, reel 56, George Bird Grinnell Papers; Charles Aubrey to George Bird Grinnell, 1 June 1908, ibid.

26. "The Bison Comes to Its Own," *NYT*, 28 May 1908; WTH, "Subscriptions to the Montana Natural Bison Herd Fund by State, to and Including January 9, 1909," box 4, American Bison Society Papers, Wildlife Conservation Society; WTH, *Second Annual Report of the American Bison Society, 1908-1909*, copy in box 4, p. 13, ibid.; WTH to Palmer H. Langdon, 4 March 1911, Outgoing Correspondence, reel 1, WTHP-WCS.

27. American Bison Society to Estate C. E. Conrad, 14 September 1909, box 4, American Bison Society Papers, Wildlife Conservation Society; WTH to James Wilson, 14 December 1909, box 4, ibid.; WTH, *Thirty Years War for Wild Life*, 169; WTH, *Our Vanishing Wild Life*, 180; WTH, "Eighty Fascinating Years," chap. 10:6.

28. Zontek, *Buffalo Nation*.

29. WTH, *Thirty Years War for Wild Life*, 173; Busch, *The War Against the Seals*; Dorsey, *The Dawn of Conservation Diplomacy*.

30. WTH to Henry Wood Elliott, 1 November 1909, Personal Letter Book, vol. 7, box 79, WTHP-LC.

31. WTH to Henry Wood Elliott, 5 November 1909, Personal Letter Book, vol. 7, box 79, WTHP-LC; WTH to Henry Wood Elliott, 8 November 1909, ibid.

32. "Wants Seals Protected," *NYT*, 27 February 1910; Charles Nagel to George Bowers, 29 April 1910, U.S. Congress, House, *Seal Islands of Alaska* (Washington, D.C.: GPO, 1911), 941.

33. Letter to Henry Wood Elliott quoted in WTH, *Thirty Years War for Wild Life*, 180; letter to Nagel quoted in WTH, "A Square Deal for the Fur Seal," box 3, WTHP-WCS.

34. Letter to Nagel quoted in WTH, "A Square Deal for the Fur Seal," box 3, WTHP-WCS; WTH to Henry Fairfield Osborn, 3 August 1910, Personal Letter Book, vol. 8, box 79, WTHP-LC.

35. Joseph Dixon to WTH, 11 January 1911, Scrapbook, 180; *Alaskan Seal Fisheries*, 4-5, 8; "The Slaughter of Baby Seals," *NYT*, 15 June 1911.

36. WTH to Henry Wood Elliott, 31 October 1910, Personal Letter Book, vol. 7,

box 79, WTHP-LC; WTH to Edmund Seymour, 8 June 1911, Outgoing Correspondence, reel 1, WTHP-WCS.

37. WTH to Madison Grant, 2 February 1911, Outgoing Correspondence, reel 1, WTHP-WCS; Graham, *Man's Dominion*, 191; G. B. Grinnell, "A Plank"; WTH, *Thirty Years War for Wild Life*, 155 (quote), 157 (facts on game meat).

38. WTH to Dr. Honsinger, 16 March 1911, Outgoing Correspondence, reel 1, WTHP-WCS.

39. *Daily North Side News*, 16 March 1911, in Scrapbook, vol. 3:5; G. B. Grinnell, "A Plank," 89; WTH, "Forbid the Sale of Game," *Evening Mail*, 13 March 1911, in Scrapbook, vol. 3.

40. Henry Fairfield Osborn to WTH, 5 April 1911, box 55, WTHP-LC; WTH to George O. Shields, 4 April 1911, Outgoing Correspondence, reel 1, WTHP-WCS.

41. WTH to Lawrence Trowbridge, 7 April 1911, Outgoing Correspondence, reel 1, WTHP-WCS; WTH to Howard Bayne, 14 April 1911, ibid.; *Lockport Union-Sun*, 8 April 1911, in Scrapbook, vol. 3:9; WTH to Franklin Hooper, 27 April 1911, reel 1, Outgoing Correspondence, WTHP-WCS; WTH to Howard Bayne, 3 May 1911, ibid.; WTH to Walter Wood, 19 May 1911, ibid.; WTH to Howard Bayne, 19 May 1911, ibid.; Howard Bayne to WTH, 6 June 1911, ibid.; WTH, *Thirty Years War for Wild Life*, 159.

42. WTH to Theodore Roosevelt, 28 May 1911, Theodore Roosevelt Papers; Burnham, "What Are the Facts?"

43. "The Bayne Bill," *Bird-Lore* 13 (July–August 1911): 225; Graham, *The Audubon Ark*, 80; Warren, *The Hunter's Game*, 55–56.

44. WTH, *Thirty Years War for Wildlife*, 155, 156–57; WTH to Palmer H. Langdon, 14 March 1911, Outgoing Correspondence, reel 1, WTHP-WCS.

45. WTH to Henry Clay Frick, 18 May 1913, Outgoing Correspondence, reel 2, ibid.

46. WTH to William Coffin, 4 March 1912, Outgoing Correspondence, reel 1, ibid. Hornaday's financial statements are in box 32, WTHP-LC.

6. OUR VANISHING WILDLIFE

1. WTH to A. H. Fox, 27 March 1911, Outgoing Correspondence, reel 1, WTHP-WCS; WTH to John M. Phillips, 20 March 1911, Personal Letter Book, vol. 9, box 79, WTHP-LC; WTH, "The Loss of a Great Opportunity," 83–86.

2. WTH to A. H. Fox, 27 March 1911, Outgoing Correspondence, reel 1, WTHP-WCS.

3. William Dutcher to WTH, 26 February 1906, quoted in WTH, *Our Vanishing Wild Life*, 151; "Minutes of the Board of Directors of the National Association of Audubon Societies, June 2, 1911," box A-174, National Association of Audubon Society Papers.

4. WTH to John M. Phillips, 3 June 1911, Outgoing Correspondence, reel 1, WTHP-WCS; "W. T. Hornaday Denounces $25,000 Gift by Gun Makers to Audubon Society," *New York Herald*, 3 June 1911; George Bird Grinnell to Joel Allen, 14 June 1911, reel 15, George Bird Grinnell Papers; WTH to Henry Fairfield Osborn,

19 May 1911, Outgoing Correspondence, reel 1, WTHP-WCS; "Never Tried to Control Society, Says Gun Manufacturer," *New York Herald*, 4 June 1911.

5. Thomas Gilbert Pearson to Joel Allen, 14 June 1911, box A-175, National Association of Audubon Society Papers; Thomas Gilbert Pearson to Frank Chapman, 15 June 1911, box A-185, ibid.

6. George Bird Grinnell to William A. Brewster, 19 June 1911, reel 15, George Bird Grinnell Papers; George Bird Grinnell to Madison Grant, 30 June 1911, ibid.

7. Pearson, *Adventures in Bird Protection*, 233.

8. See WTH to A. S. Houghton, 8 June 1911, Outgoing Correspondence, reel 1, WTHP-WCS.

9. John Bird Burnham to Commissioner Whipple, 21 September 1911, box 4, John Bird Burnham Papers.

10. John Bird Burnham quoted in DeSormo, *John Bird Burnham*, 192; WTH to John Bird Burnham, 10 January 1911, box 4, John Bird Burnham Papers; WTH to John Bird Burnham, 19 October 1911, Outgoing Correspondence, reel 1, WTHP-WCS; WTH to Henry Fairfield Osborn, 25 September 1911, ibid.; WTH to Madison Grant, 28 October 1911, ibid.

11. WTH to John Dix, 22 November 1911, Outgoing Correspondence, reel 1, WTHP-WCS; WTH to Madison Grant, 18 December 1911, ibid.; WTH, "Protection of Game for the Gun Makers," *NYT*, 20 December 1911; George Bird Grinnell to John Bird Burnham, 21 December 1911, reel 15, George Bird Grinnell Papers; John Bird Burnham, "Mr. Burnham Retorts," *NYT*, 16 December 1911; John Bird Burnham, "Opposes Hornaday's Views," *NYT*, 21 December 1911; WTH to Theodore Roosevelt, 9 December 1911, Theodore Roosevelt Papers.

12. WTH, *Tales from Nature's Wonderlands*, 108-9.

13. WTH to Henry Fairfield Osborn, 20 January 1912, Outgoing Correspondence, reel 1, WTHP-WCS; George Bird Grinnell to Madison Grant, 7 February 1912, reel 17, George Bird Grinnell Papers; WTH, "Eighty Fascinating Years," chap. 20:12; WTH to George O. Shields, 31 October 1912, Outgoing Correspondence, vol. 3, WTHP-WCS.

14. For a fuller appreciation of the landmark 1911 convention, see Dorsey, *The Dawn of Conservation Diplomacy*, 159-64.

15. Clark, "Recent Fur Seal Legislation"; Townsend, "The Pribilof Fur Seal Herd"; Dorsey, *The Dawn of Conservation Diplomacy*, 109-12; Henry Fairfield Osborn to William Sulzer, 22 January 1912, quoted in Elliott, "A Statement Submitted in Re the Fur Seal Herd of Alaska," 205-6; WTH to Henry Wood Elliott, 27 January 1911 (really 1912), Outgoing Correspondence, reel 1, WTHP-WCS.

16. U.S. Congress (Senate), *Alaska Seal Fisheries*, 8; WTH to Henry Wood Elliott, 17 February 1912, Personal Letter Book, vol. 10, box 80, WTHP-LC.

17. Jordan, *The Days of a Man*, 1:611, 609; WTH, "The Rescued Fur Seal Industry," 81; Busch, *The War Against the Seals*, 97.

18. WTH to George A. Clark, 5 February 1913, Outgoing Correspondence, vol. 4, WTHP-WCS; George A. Clark to WTH, 12 February 1913, Scrapbook, vol. 6; "Clark Defends His Seal Report," *NYT*, 21 February 1914.

19. WTH to George O. Shields, 1 February 1912, Outgoing Correspondence,

reel 1, WTHP-WCS; WTH to John Weeks, 9 September 1912, ibid.; WTH to Madison Grant, 12 September 1912, ibid.

20. WTH, *Thirty Years War for Wild Life*, 162.

21. WTH to William Coffin, 13 September 1912, Outgoing Correspondence, reel 2, WTHP-WCS; DeSormo, *John Bird Burnham*, 193.

22. WTH to Austin Lutchaw, 12 September 1912, Outgoing Correspondence, reel 2, WTHP-WCS; WTH, *Thirty Years War for Wild Life*, 162; WTH, "Wild Life Protection," 347, 345.

23. WTH to Henry Fairfield Osborn, 10 September 1912, Personal Letter Book, vol. 11, box 80, WTHP-LC; WTH to Henry Shoemaker, 13 September 1912, Outgoing Correspondence, reel 2, WTHP-WCS.

24. Quoted in Fox, *John Muir and His Legacy*, 150.

25. G. G., "The Distribution of Our Campaign Book," *New York Zoological Society Bulletin* 16 (May 1913): 988–89; WTH to Theodore Roosevelt, 3 February 1913, Theodore Roosevelt Papers.

26. Roosevelt, "Our Vanishing Wild Life," in *Literary Essays*, 562; WTH, *Our Vanishing Wild Life*, 56; Aldo Leopold quoted in Meine, *Aldo Leopold*, 150.

27. George Gladden, "A Champion of Wild Life," *American Review of Reviews* 48 (December 1913): 698; Meine, *Aldo Leopold*, 128. Leopold's biographers do not all agree on the extent of Hornaday's influence. Julianne Lutz Newton argues that Hornaday undoubtedly influenced Leopold, but not to the degree that Meine claims (see Newton, *Aldo Leopold's Odyssey*, 91). Susan Flader greatly downplays Hornaday's impact (see Flader, *Thinking Like a Mountain*, 12).

28. George Bird Grinnell to John Bird Burnham, 23 January 1913, reel 16, George Bird Grinnell Papers; John Bird Burnham to George Bird Grinnell, 19 February 1913, ibid.; DeSormo, *John Bird Burnham*, 180–81.

29. Thomas Gilbert Pearson, "The Weeks-McLean Act," *Bird-Lore* 14 (March–April 1913): 138; John Bird Burnham to George Bird Grinnell, 12 March 1913, George Bird Grinnell Papers.

30. Pearson, *Adventures in Bird Protection*, 278; WTH, *Our Vanishing Wild Life*, 250; WTH to George O. Shields, 25 April 1913, Outgoing Correspondence, vol. 4, WTHP-WCS.

31. WTH to Henry Fairfield Osborn, 10 June 1913, Outgoing Correspondence, vol. 5, WTHP-WCS.

32. Pearson, *Adventures in Bird Protection*, 279; WTH to David Houston, 27 September 1913, Outgoing Correspondence, vol. 6, WTHP-WCS.

33. Graham, *Man's Dominion*, 47; John Burroughs, "To the Editors," *Forest & Stream* 26 (4 March 1886): 104; Charles Dudley Warner, "To the Editors," *Forest & Stream* 26 (15 April 1886): 222; WTH, "Destruction of Our Birds and Mammals," 87; McIver, *Death in the Everglades*.

34. Pearson, *Adventures in Bird Protection*, 260–61.

35. WTH to Theodore Roosevelt, 3 February 1913, Theodore Roosevelt Papers; Pearson, *Adventures in Bird Protection*, 261–62; U.S. House Committee on Ways and Means, *Tariff Schedules*, 62nd Cong., 3rd sess., vol. 5 (Washington, D.C.: GPO, 1913), 5308, 5311.

36. U.S. House Committee on Ways and Means, *Tariff Schedules*, 5330, 5344–45.

37. WTH to Theodore Roosevelt, 2 February 1913, Theodore Roosevelt Papers; Pearson, *Adventures in Bird Protection*, 262; Francis Harrison to WTH, 11 March 1913, box 57, WTHP-LC; WTH to Miss M. S. Davies, 22 April 1913, Outgoing Correspondence, vol. 4, WTHP-WCS.

38. WTH to Madison Grant, 28 April 1913, Outgoing Correspondence, vol. 4, WTHP-WCS; Madison Grant to WTH, 2 May 1913, ibid.

39. U.S. Senate Subcommittees of the Committee on Finance, *Tariff Schedules*, 63rd Cong., 1st sess. (Washington, D.C.: GPO, 1913), 706.

40. Ibid.; WTH, "Foreign Wild Birds," *NYT*, 5 May 1913; Pearson, *Adventures in Bird Protection*, 263; Thomas Gilbert Pearson, "The Feathers Proviso," *Bird-Lore* 14 (July–August 1913): 272–73; WTH to Nina Hornaday, 26 June 1913, Outgoing Correspondence, reel 3, WTHP-WCS; "Persistent War on Birds," *NYT*, 21 June 1913; WTH to Henry Fairfield Osborn, 10 July 1913, Outgoing Correspondence, reel 3, WTHP-WCS.

41. "Say Feather Men Dominate Senate," *NYT*, 3 August 1913; WTH to Henry Oldys, 29 July 1913, box 5, WTHP-WCS; Pearson, *Adventures in Bird Protection*, 264–65.

42. Woodrow Wilson to Edith A. Wilson, 7 August 1913, in Wilson, *The Letters of Woodrow Wilson*, 28:129.

43. George Bird Grinnell to Thomas Gilbert Pearson, 17 September 1913, box A-68, National Association of Audubon Society Papers.

44. Henry Oldys to WTH, 15 September 1913, box 5, WTHP-WCS; WTH to Henry Oldys, 18 September 1913, ibid.; WTH to Thomas Gilbert Pearson, 22 September 1913, A-68, National Association of Audubon Society Papers; Thomas Gilbert Pearson to WTH, 24 September 1913, ibid.

45. WTH to Nina Hornaday, 14 April 1914, Outgoing Correspondence, reel 3, WTHP-WCS; Madison Grant, "France Awards a Medal for Bird Protection," *Permanent Wildlife Protection Fund Bulletin* 1 (1915): 89–92; "France Honors WT Hornaday," *New York Tribune*, 25 March 1914, www.loc.gov.

46. WTH to Mr. Richard Hargrave, 12 December 1913, Outgoing Correspondence, vol. 7, WTHP-WCS; "Carmody Bitterly Attacks Hornaday," *NYT*, 22 December 1913.

7. THE GREAT WAR

1. WTH to Thomas Gilbert Pearson, 21 January 1914, Outgoing Correspondence, vol. 7, WTHP-WCS; WTH to JCH, 17 February 1914, box 1, WTHP-LC.

2. WTH, "The Seamy Side of the Protection of Wild Game," *NYT Magazine*, 8 March 1914, 3.

3. WTH to Madison Grant, 2 May 1913, Outgoing Correspondence, vol. 4, WTHP-WCS; WTH to James O'Gorman, 24 April 1914, Outgoing Correspondence, vol. 7, ibid.; "Stop Killing Useful Birds, He Says," *Washington Post*, 4 May 1914, www.ancestry.com.

4. *Congressional Record*, 63rd Cong., 2nd sess., 1914, 51, pt. 9: 9109, 9107–13; WTH to Theodore Roosevelt, 29 May 1914, Theodore Roosevelt Papers.

5. WTH, *Wild Life Conservation in Theory and Practice*, v; Charles C. Adams, "Zoologists, Teachers and Wildlife Conservation," *Science* 41 (28 May 1915): 791.

6. Hays, *Conservation and the Gospel of Efficiency*, 40.

7. "Extract from the Minutes of the Annual Meeting of the NYZS held on January 12, 1915," in Scrapbook, vol. 7:5, WTHP-WCS; "Plea for Moose by Mr. Roosevelt," *New York Herald*, 13 January 1915, copy in Scrapbook, vol. 7; "Forest Reserves as Game Refuges," *Forest & Stream* 58 (8 February 1902): 101.

8. WTH, "Basic Requirements for National Forest Sanctuaries," in Scrapbook, vol. 7, WTHP-WCS.

9. WTH, "Belgium," in *Old Fashioned Verses*, 3–4; WTH, "Eight Fascinating Years," chap. 19:3.

10. WTH, "Robbed," in *Old Fashioned Verses*, 67; itinerary culled from newspaper accounts in Scrapbook, vol. 7, WTHP-WCS.

11. Meine, *Aldo Leopold*, 149; WTH, *Thirty Years War for Wild Life*, 192.

12. WTH to Henry Graves, 21 October 1915, Outgoing Correspondence, vol. 10, WTHP-WCS; WTH, *Thirty Years War for Wild Life*, 215; George Hunt to WTH, 1 September 1915, Scrapbook, vol. 7, WTHP-WCS.

13. William Spry to WTH, 13 November 1915, Scrapbook, vol. 7, WTHP-WCS; WTH to William Spry, ibid.; "Hornaday Plan Causes Discussion," *Salt Lake Times*, 21 November 1915, copy ibid.

14. WTH to Henry Wood Elliott, 30 December 1915, Outgoing Correspondence, reel 5, WTHP-WCS.

15. Edwin Nelson to WTH, 24 December 1915, Scrapbook, vol. 7, WTHP-WCS; WTH to Carl Hayden, 31 January 1916, Outgoing Correspondence, reel 5, ibid.; WTH to Carl Hayden, 2 February 1916, ibid.

16. WTH to Theodore S. Palmer, 14 February 1916, Outgoing Correspondence, reel 5, WTHP-WCS; WTH to Theodore S. Palmer, 14 January 1916, ibid.

17. Burgess, *Now I Remember*, 131; WTH, *Thirty Years War for Wild Life*, 192.

18. WTH to Reed Smoot, 27 April 1916, Outgoing Correspondence, reel 6, WTHP-WCS; WTH to George Hewitt Myers, 21 September 1916, reel 7, ibid.; WTH to Theodore Roosevelt, 16 December 1916, Theodore Roosevelt Papers; WTH to David Miller, 23 March 1917, WTHP-GT.

19. WTH, *Thirty Years War for Wild Life*, 213.

20. WTH to Aldo Leopold, 8 January 1916, Outgoing Correspondence, reel 6, WTHP-WCS.

21. WTH, "Ethics in Hunting Game," 103; WTH to Aldo Leopold, 24 August 1917, reel 8, WTHP-WCS; see also WTH to Theodore S. Palmer, 24 August 1917, ibid.; WTH to Aldo Leopold, 2 October 1917, ibid.; "The Position of the Audubon Society," *Forest & Stream* 87 (September 1917): 406.

22. WTH, "The Duty of the Citizen in Our Navy," 17 December 1916, box 86, WTHP-LC; WTH to G. W. Lemon, 14 May 1917, Outgoing Correspondence, United States Junior Naval Reserve (USJNR), WTHP-WCS; WTH, "Eighty Fascinating Years," chap. 20:7; Josephus Daniels to Joseph P. Tumulty, 12 December 1918, Woodrow Wilson Papers, Library of Congress; WTH to Victor Stockell, 1 February 1917, Outgoing Correspondence, USJNR, WTHP-WCS.

23. WTH to David Miller, 23 March 1917, WTHP-GT; WTH to Edmund Sey-

mour, 27 December 1916, box 9, American Bison Society Papers, Denver Public Library.

24. WTH to David Miller, 23 March 1917, WTHP-GT; WTH to Henry Wood Elliott, 2 July 1917, Outgoing Correspondence, reel 8, WTHP-WCS; WTH to David Miller, 23 March 1917, WTHP-GT.

25. WTH, *Awake! America*, xii; WTH to George MacDonald, 7 May 1917, Outgoing Correspondence, reel 8, WTHP-WCS; WTH to David Miller, 29 April 1918, WTHP-GT.

26. Bridges, *Gathering of Animals*, 371, 368; WTH to Aldo Leopold, 23 April 1917, Outgoing Correspondence, reel, 7, WTHP-WCS; WTH to Charles Farrelly, 20 September 1919, reel 9, ibid.

27. WTH, *Awake! America*, 196; "Attack Disloyalty at Teacher's Rally," *NYT*, 28 November 1917; Theodore Roosevelt to WTH, 25 November 1917, in *The Letters of Theodore Roosevelt*, 7:1252; WTH, *Awake! America*, 3, 7–11.

28. WTH to Lionel Sutro, 1 February 1919, Outgoing Correspondence, reel 9, WTHP-WCS; WTH to David Miller, 15 May 1918, WTHP-GT; WTH to Dr. W. Dreyser, 30 September 1919, Outgoing Correspondence, reel 9, WTHP-WCS; WTH to David Miller, 3 June 1919, WTHP-GT.

29. WTH to Alfred E. Smith, 14 November 1918, Personal Letter Book, vol. 18, box 82, WTHP-LC; WTH to Fred D'Osta, 29 October 1919, Outgoing Correspondence, reel 9, WTHP-WCS.

30. WTH, *The Lying Lure of Bolshevism*, 38; Henry L. Mencken to Fielding Hudson Garrison, 26 December 1919, in *The Letters of H. L. Mencken*, 166–67; WTH to James Henry Rice, 22 January 1920, Outgoing Correspondence, reel 9, WTHP-WCS.

31. WTH to David Miller, 3 June 1919, 23 March 1917, 15 May 1918, 3 June 1919, WTHP-GT; WTH to Charles Sackett, 8 October 1919, Outgoing Correspondence, reel 9, WTHP-WCS.

32. WTH to David Miller, 15 May 1918, WTHP-GT.

33. Graham, *Man's Dominion*, 202; WTH to Henry Wood Elliott, 20 February 1918, Outgoing Correspondence, reel 8, WTHP-WCS.

34. George Bird Grinnell to Theodore Roosevelt, 28 May 1894, reel 4, George Bird Grinnell Papers; George Bird Grinnell to Charles Sheldon, 23 February 1918, reel 40, ibid.; Charles Sheldon to George Bird Grinnell, 14 March 1918, ibid.

35. WTH to David Houston, 16 February 1918, Outgoing Correspondence, reel 8, WTHP-WCS; WTH to Edwin Nelson, 2 March 1918, copy in reel 40, George Bird Grinnell Papers; WTH to Madison Grant, 7 March 1918, Outgoing Correspondence, reel 8, WTHP-WCS; WTH to Edwin Nelson, 11 March 1918, ibid.

36. WTH to Wright Wenrich, 12 April 1918, Outgoing Correspondence, reel 8, WTHP-WCS; WTH to Edwin Nelson, 15 March 1921, ibid., reel 10.

37. WTH, *Awake! America*, 57.

38. WTH, *Thirty Years War for Wild Life*, 190, 197–198; WTH, "Two Great Campaigns for Wildlife Sanctuaries," 72; WTH to Clark Williams and A. Barton Hepburn, 1 March 1920, Outgoing Correspondence, reel 9, WTHP-WCS.

39. WTH, "Conservation! The Great Task for Boys," 36; WTH, "The Permanent Fund Establishes a Medal"; Macleod, *Building Character in the American Boy*,

246. For the current status of the Hornaday award, see www.scouting.org/scout source/Awards/HornadayAwards.aspx.

8. FIGHTING THE ESTABLISHMENT

1. WTH to Edmund Seymour, 28 February 1920, box 9, American Bison Society Papers, Denver Public Library; "Calls Dr. Hornaday 'Autocrat of the Zoo,'" *NYT,* 15 March 1920

2. "Calls Dr. Hornaday 'Autocrat of the Zoo,'" *NYT,* 15 March 1920; WTH to John M. Phillips, 19 April 1920, Outgoing Correspondence, WTHP-WCS; WTH to William Sulzer, 17 March 1920, ibid.; WTH to Henry Wood Elliott, 7 March 1920, ibid.; WTH to William Ford, 30 June 1920, ibid.; "Hirschfield Renews Hornaday Attack," *NYT,* 1 April 1920.

3. Thomas Riggs to George Bird Grinnell, 4 March 1920, reel 40, George Bird Grinnell Papers; George Bird Grinnell to Thomas Riggs, 9 March 1920, ibid.; Thomas Riggs to George Bird Grinnell, 3 June 1920, ibid.

4. WTH, "Who Is Responsible for the Exterminative Programs?," 135; WTH to Edmund Seymour, 28 February 1920, box 9, American Bison Society Papers, Denver Public Library; WTH to Henry Fairfield Osborn, 20 January 1921, Outgoing Correspondence, reel 10, WTHP-WCS; WTH to George Bird Grinnell, 9 August 1920, reel 40, George Bird Grinnell Papers; WTH to Marshall McLean, 18 September 1920, Outgoing Correspondence, reel 10, WTHP-WCS.

5. WTH to Edwin Nelson, 15 December 1920, Outgoing Correspondence, reel 10, WTHP-WCS; Sherwood, *Big Game in Alaska,* 41.

6. Gabrielson, *Wildlife Refuges,* 12–13.

7. WTH to Edward Howe Forbush, 9 January 1922, Outgoing Correspondence, reel 10, WTHP-WCS; WTH to Aldo Leopold, 1 July 1920, ibid.; WTH to Edmund Seymour, 8 August 1921, ibid.

8. WTH to W. D. McLaughlin, 12 July 1920, Outgoing Correspondence, reel 10, WTHP-WCS; *Bird-Lore* 24 (May–June 1922): 164.

9. WTH, *Hornaday's American Natural History,* vii; Carl Van Doren, "Books: The Roving Critic," *Nation* 116 (11 October 1922): 377; WTH, *The Minds and Manners of Wild Animals,* 286–87; *Science* 56 (24 November 1922): 605. A more competent study contemporary to *Minds and Manners* is Holmes, *The Evolution of Animal Intelligence.*

10. WTH to John Bird Burnham, 19 December 1922, Outgoing Correspondence, reel 11, WTHP-WCS; "Minutes of the Annual Meeting of the Advisory Board, Dec 12, 1923," Scrapbook, vol. 9, WTHP-WCS.

11. "Minutes of the Annual Meeting of the Advisory Board, Dec 12, 1923," Scrapbook, vol. 9, WTHP-WCS.

12. John Bird Burnham to Galvin, 27 June 1924, quoted in *Congressional Record,* 69th Cong., 1st sess., 20.

13. Will Dilg quoted in "The Vanishing Wild Life of America," *Literary Digest* 74 (9 September 1922): 60; Scarpino, *Great River,* 46–47.

14. WTH to Edwin Nelson, 24 June 1924, Outgoing Correspondence, reel 13, WTHP-WCS; WTH to Henry Wallace, 15 June 1924, ibid.; WTH to Henry Fair-

field Osborn, 3 July 1924, ibid.; WTH to Henry Fairfield Osborn, 30 June 1924, ibid.; WTH to Madison Grant, 10 July 1924, ibid.; WTH to Will Dilg, 6 November 1924, ibid.

15. WTH to Captain A. E. Burghduff, 8 December 1924, Outgoing Correspondence, reel 13, WTHP-WCS; WTH to Edwin Nelson, 2 December 1924, ibid.; Edwin Nelson quoted in WTH, *Thirty Years War for Wild Life,* 125; WTH to Clark Williams, 8 December 1924, Outgoing Correspondence, reel 13, WTHP-WCS.

16. WTH to John A. McGuire, 27 February 1925, Outgoing Correspondence, reel 14, WTHP-WCS; WTH to John Bird Burnham, 16 May 1925, ibid.; WTH to Will Dilg, 26 May 1925, Outgoing Correspondence, reel 14, WTHP-WCS; WTH to Madison Grant, 25 May 1925, ibid.

17. WTH, "Transcontinental Observations by the Author, July 1925," Scrapbook, vol. 9, WTHP-WCS.

18. Nelson's report is summarized in "The Government Report on Suggestion that Bag-Limits Be Reduced," *Bird-Lore* 27 (July–August 1925): 300–301.

19. WTH, *Thirty Years War for Wild Life,* 125–26; "Says Arms Men Foil Game Conservation in Federal Bureau," *NYT,* 12 August 1925; *Nashville Banner,* quoted in "Our Game Protectors at War," *Literary Digest* 90 (5 September 1925): 16.

20. Pearson, *Adventures in Bird Protection,* 294; WTH to Will Dilg, 4 September 1925, Outgoing Correspondence, reel 15, WTHP-WCS; WTH to James Spear, 8 September 1925, ibid.

21. Pearson, *Adventures in Bird Protection,* 296–97; John Bird Burnham to W. E. Covey, 24 September 1925, box 11, John Bird Burnham Papers; WTH to Will Dilg, 16 October 1925, Outgoing Correspondence, reel 15, WTHP-WCS; WTH to John A. McGuire, 17 October 1925, ibid.; Fox, *John Muir and the American Conservation Movement,* 166.

22. William C. Adams to Thomas Gilbert Pearson, 28 September 1925, box A-126, National Association of Audubon Society Papers; George Bird Grinnell to Thomas Gilbert Pearson, 13 October 1925, A-125, ibid.; WTH to Madison Grant, 16 October 1925, Outgoing Correspondence, reel 15, WTHP-WCS; WTH to Madison Grant, 21 October 1925 and 27 November 1925, ibid.

23. Thomas Gilbert Pearson to Frank Chapman, 17 February 1926, box A-125, National Association of Audubon Society Papers; Robert Sterling Yard to John M. Phillips, 25 March 1926, reel 36, George Bird Grinnell Papers; Gustavus Pope to Thomas Gilbert Pearson, 8 February 1926, A-125, National Association of Audubon Society Papers.

24. Thomas Gilbert Pearson to Donald Campbell, 6 May 1926, box A-125, National Association of Audubon Society Papers; Schuyler Merritt to George Bird Grinnell, 20 March 1926, reel 43, George Bird Grinnell Papers; John C. Phillips to George Bird Grinnell, 26 March 1926, reel 39, ibid.

25. John Bird Burnham to Madison Grant, 18 March 1926, box 11, John Bird Burnham Papers; WTH to Madison Grant, 21 April 1926, reel 15, WTHP-WCS.

26. *Congressional Record,* 69th Cong., 1st sess., vol. 67, pt. 8: 8476, 8478; John Bird Burnham to George Lawyer, 12 May 1926, box 11, John Bird Burnham Papers; Edward Howe Forbush to Thomas Gilbert Pearson, 19 October 1926, A-125,

National Association of Audubon Society Papers; WTH to John A. McGuire, 5 May 1926, reel 15, WTHP-WCS.

27. Frederic C. Walcott to Hiram Bingham, 24 May 1926, box A-125, National Association of Audubon Society Papers; WTH to Louis Agassiz Fuertes, 21 January 1926, Outgoing Correspondence, reel 15, WTHP-WCS; WTH, "Eighty Fascinating Years," chap. 20:15–16; Hancocks, *A Different Nature*, 103–10.

28. "W. T. Hornaday Retires," *NYT*, clipping in scrapbook, WTHP-GT

29. "Dr. Hornaday's Retirement as Director of the New York Zoological Park," *Scientific Monthly* 23 (July 1926): 91; "William T. Hornaday," *NYT*, 22 May 1926; WTH, "Eighty Fascinating Years," chap. 11:16.

30. WTH to Edmund Seymour, 19 May 1932, box 2, American Bison Society Papers, Denver Public Library; WTH, *Thirty Years War for Wild Life*, 153–54.

31. Madison Grant to WTH, 26 November 1928, box 75, WTHP-LC; Lee Crandall to WTH, 30 September 1927, box 13, ibid.

32. Louis Pammel to WTH, 23 April 1926, Louis Pammel Papers; WTH to Louis Pammel, 29 April 1926, ibid.; WTH to Louis Pammel, 21 June 1926, ibid.; "To Dedicate 3 Historical Spots in Ames," *Ames Tribune*, 11 June 1926.

33. WTH, *Wild Animal Interviews*, 175; WTH to Edmund Seymour, 25 October 1926, handwritten note tucked inside the front cover of a copy of *Wild Animal Interviews* borrowed via interlibrary loan from Williams College; Charles F. Dunn to WTH, 15 February 1928, box 75, WTHP-LC; Madison Grant to WTH, 15 February 1928, ibid.

34. WTH, "Eighty Fascinating Years," chap. 20:14; WTH to Charles Davenport, 6 October 1926, Davenport Papers.

35. "Dr. W. T. Hornaday Noted Naturalist, Dies at Home," *Stamford Advocate*, 8 March 1937; Shipley, "A Study of Conservation Philosophies," 188, 1–3.

36. Edwin Nelson to J. Sanford Barnes, 18 December 1926, reel 43, George Bird Grinnell Papers; WTH, *Thirty Years War for Wild Life*, 127.

37. Edwin Nelson to Members of the Migratory Bird Treaty Act Advisory Board, 8 March 1927, reel 43, George Bird Grinnell Papers; WTH, "The Ducks and Geese Lose Again," press release, Permanent Wildlife Protection Fund, 26 May 1927, copy in Scrapbook, vol. 9, WTHP-WCS.

38. "Dr. W. T. Hornaday Talks on Career at Convocation," *Iowa State Student*, 30 April 1927; "Hornaday Will Speak Tomorrow at Convocation," *Iowa State Student*, 28 April 1927.

39. "Libel Action Filed Against Hornaday," *NYT*, 17 July 1927; John Bird Burnham to John D. Burnham, 10 August 1926, box 11, John Bird Burnham Papers; John Bird Burnham to Edward Baetjer, 22 August 1927, George Bird Grinnell Papers.

40. *New York Misc. Reports*, 130, Misc. 207 (1927), 223 New York State 750, www.loislaw.com; "Hornaday Loses Point in Burnham Libel Suit," *NYT*, 26 August 1927; John Bird Burnham to Edward Baetjer, 22 August 1927, George Bird Grinnell Papers; Walter Pollack to John Bird Burnham, 6 June 1929, box 2, John Bird Burnham Papers.

41. WTH, *Thirty Years War for Wild Life*, 235; "Mr. Hornaday's Pessimism,"

Columbus Evening Dispatch, December 1927, copy in Scrapbook, vol. 9, WTHP-WCS.

9. FIGHTING TO THE END

1. WTH to Peter Norbeck, 19 January 1928, quoted in WTH, *Thirty Years War for Wild Life,* 236; Peter Norbeck to WTH, 21 January 1928, Scrapbook, vol. 10, WTHP-WCS.
2. Carlos Avery to Peter Norbeck, 12 April 1928, Peter Norbeck Papers; Peter Norbeck to Edmund Seymour, 10 April 1928, ibid.
3. WTH, *Thirty Years War for Wild Life,* 239; Peter Norbeck to Edmund Seymour, 21 April 1928, Peter Norbeck Papers.
4. WTH, "The Final Test of American Game Conservationists," circular memo, 25 November 1928, copy in Peter Norbeck Papers; WTH, "American Poor Game Protectors," *Plain Truth,* 1 December 1928; "Game Refuge Bill Urged by Hornaday," *NYT,* 1 December 1928; WTH to Peter Norbeck, 16 January 1929, Peter Norbeck Papers; Brant, *Adventures in Conservation,* 5.
5. Pearson, *Adventures in Bird Protection,* 302; Thomas Gilbert Pearson to George Bird Grinnell, 29 January 1929, reel 43, George Bird Grinnell Papers; Thornton Burgess to WTH, 6 February 1929, box 75, WTHP-LC; Hayden, *The International Protection of Wildlife,* 84.
6. Copy of poem in box 1, WTHP-LC; Helen Hornaday Fielding to Friends, 28 August 1929; copy in scrapbook, WTHP-GT; Isabel Ross, "Hornaday Wed 50 Years: Pays Tribute to Wife," *New York Tribune,* 12 September 1929, ibid.
7. Van Name, Miller, and Quinn, *A Crisis in Conservation,* 14; WTH to Irving Brant, 8 July 1929, box 17, Irving Brant Papers; WTH to Davis Quinn, 29 November 1929, ibid.
8. Furmansky, *Rosalie Edge,* 88–90; Rosalie Edge, "Autobiography," manuscript, copyright Peter Edge, 15–23.
9. R. Edge, "Autobiography," 19, 20; Peter Edge to author, 25 October 1999.
10. R. Edge, "Autobiography," 27; Brant, *Adventures in Conservation,* 17; WTH to Rosalie Edge, 24 July 1931, box 17, Irving Brant Papers; R. Edge, "Autobiography," 26.
11. WTH to Irving Brant, 14 November 1929, box 17, Irving Brant Papers; George Bird Grinnell to I. T. Quinn, 5 November 1929, reel 42, George Bird Grinnell Papers.
12. Harry McGuire to Arthur Holhaus, 6 January 1930, box 2, American Bison Society Papers, Denver Public Library; WTH, *Thirty Years War for Wild Life,* 130; WTH to Gilbert Haugen, 4 January 1930, box 2, American Bison Society Papers, Denver Public Library; WTH to Irving Brant, 8 January 1930, box 17, Irving Brant Papers; Irving Brant to WTH, 24 February 1930, box 2, American Bison Society Papers, Denver Public Library; WTH to Edmund Seymour, 15 March 1930, ibid.
13. Brant, *Compromised Conservation;* R. Edge, "Autobiography," 35; WTH to Edmund Seymour, 7 October 1930, box 2, American Bison Society Papers, Denver Public Library; Brant, *Adventures in Conservation,* 18; Irving Brant to Robert

Cushman Murphy, 1 December 1932, Rosalie Edge Papers, Denver Public Library; Hornaday quoted in Brant, *Adventures in Conservation*, 18.

14. WTH, *Thirty Years War for Wild Life*, 139; WTH to Robert Cushman Murphy, 23 October 1930, Robert Cushman Murphy Papers, American Philosophical Society, Philadelphia; WTH to Edmund Seymour, 29 October 1930, box 2, American Bison Society Papers, Denver Public Library; WTH, *Thirty Years War for Wild Life*, 140; "The Twenty-Sixth Annual Meeting," *Bird-Lore* 32 (November–December 1930): 462; Graham, *The Audubon Ark*, 114.

15. R. Edge, "Autobiography," 38; Furmansky, *Rosalie Edge*, 114–66.

16. George Pratt to Rosalie Edge, 20 October 1931, Rosalie Edge Papers, Hawk Mountain; Rosalie Edge to George Pratt, 30 October 1931, ibid.; WTH to Rosalie Edge, 11 November 1931, box 11, WTHP-LC; Rosalie Edge to WTH, 18 August 1932, ibid.

17. Irving Brant to Rosalie Edge, 3 February 1931, box 17, Irving Brant Papers; Peter Edge, telephone conversation with author, 13 October 1999; Dunlap, *Saving America's Wildlife*, 96.

18. WTH to Rosalie Edge, 24 July 1931, box 17, Irving Brant Papers; WTH to Irving Brant, 24 July 1931, ibid.; Irving Brant to WTH, 17 July 1931, ibid.; Thomas Gilbert Pearson to Frank Chapman, 3 November 1931, box A-185, National Association of Audubon Society Papers; Brant, *Shotgun Conservation*.

19. Daniel Carter Beard to WTH, 131 January 1931, box 11, WTHP-LC; WTH, *Thirty Years War for Wild Life*, 12, 6.

20. WTH, *Thirty Years War for Wild Life*, vii, 247, 2.

21. Sobel, *Coolidge*, 340–41; Hoover, "Statement Announcing the Proposal of the Moratorium on Intergovernmental Debts," 326; WTH to Willis K. Miller, 20 June 1931, box 2, Michael H. Miller Collection; WTH to Herbert Hoover, 8 July 1931, Herbert Hoover Papers.

22. "Drought Depletes Game Bird Flocks," *NYT*, 4 July 1931; WTH to Herbert Hoover, 8 July 1931, Herbert Hoover Papers; R. W. Dunlap to Herbert Hoover, 10 July 1931, ibid.; Hoover, "Conservation of our Waterfowl," 156–57; WTH, "The Waterfowl Crisis," *Plain Truth*, 1 December 1931.

23. WTH, "The Scourge of Guns and Commercialism," *Plain Truth*, 1 December 1931.

24. WTH, "A Call to Action," 19 March 1932, copy in Herbert Hoover Papers.

25. WTH to Edmund Seymour, 27 April 1932, box 2, American Bison Society Papers, Denver Public Library; Brant, *A Last Plea for Waterfowl*, 8; Donald Stillman, "A New Deal for Our Wild Fowl," *Literary Digest* 116 (7 October 1933): 30.

26. WTH to Edmund Seymour, 29 July 1932, box 2, American Bison Society Papers, Denver Public Library; WTH to Edmund Seymour, 13 October 1932, ibid.; WTH to Edmund Seymour, 11 November 1932, ibid.

27. Box 15, WTHP-LC.

28. Charles F. Dunn to WTH, 20 April 1935, box 75, WTHP-LC; WTH to Charles F. Dunn, 25 April 1935, ibid.; Charles F. Dunn to WTH, 27 April 1935, ibid.; WTH to Frederick A. Stokes and Co., 4 October 1935, ibid.

29. WTH, "Eighty Fascinating Years," chap. 20:1, 4, 3.

30. WTH, "The Slump in American Morale," in "Eighty Fascinating Years," 1, 2, 3; WTH, "A Reduction of Crime in New York City," 10 December 1925, WTHP-WCS.

31. A copy of the advertisement is in WTHP-GT.

32. WTH, "Our Migratory Game," *NYT*, 13 August 1933; "Clash Marks Hearing on Wild Fowl Baiting," *NYT*, 29 August 1933.

33. WTH to Edmund Seymour, 1 December 1933, American Bison Society Papers, Denver Public Library; WTH, "Breeding Grounds and the Duck Stamp Tax," *Plain Truth*, 1 December 1933; WTH to Edmund Seymour, 11 March 1935, American Bison Society Papers, Denver Public Library; WTH to E. A. Preble, 17 July 1934, box 75, WTHP-LC.

34. WTH to E. A. Preble, 25 July 1934, box 75, WTHP-LC; WTH to E. A. Preble, 7 August 1934, ibid.; Ding Darling to Edmund Seymour, 24 September 1934, box 2, American Bison Society Papers, Denver Public Library.

35. "Hoosier, 80, Fights for Wild Life Still," *Indianapolis Star*, 27 January 1935; "Dr. Hornaday, Director of N.Y. Zoo Is Militantly Active at 80," *Denver Post*, 19 December 1934; "Dr. Hornaday Is 80 Today," *New York Sun*, 1 December 1934. All copies are in WTHP-GT.

36. Ding Darling to Henry A. Wallace and Franklin D. Roosevelt, 4 February 1935, in *FDR and Conservation*, 1:347.

37. Press release dated 8 February 1935, courtesy of Mark Madison of the National Conservation Training Center Archives; WTH, "Great Game Killing Reforms in 1935," *Plain Truth*, 1 December 1935; Brant, *Adventures in Conservation*, 52.

38. WTH to Dodge and Temple Fielding, 23 February 1935, box 9, WTHP-LC; WTH to Dodge and Temple Fielding, 10 March 1936, ibid.

39. WTH to W. Reid Blair, 15 December 1936, box 1, WTHP-WCS; WTH, *Thirty Years War for Wild Life*, 194–95.

40. "Dr. W. T. Hornaday Ill in Bed, Hopes to Be about Soon," *Stamford Advocate*, 1 December 1936, WTHP-GT.

41. "Dr. William T. Hornaday Noted Naturalist, Dead," *Stamford Advocate*, 8 March 1937; "Bird and Animal Authority," *Indianapolis Star*, 7 March 1937; "Dr. Hornaday, First Head of NY Zoo, Dead," *New York Herald Tribune*, 7 March 1937; "Dr. W. T. Hornaday, Widely-Known Naturalist, Hoosier Native, Dies," *Indianapolis Star*, 7 March 1937. Copies of all articles in WTHP-GT.

42. "W. T. Hornaday Funeral Plans," *New York Sun*, 8 March 1937; "Leaders in Nature Work of Nation at Hornaday Funeral," *Stamford Advocate*, 9 March 1937; "Notables Attend Hornaday Rites," *NYT*, 10 March 1937; "200 Employees at Funeral of Dr. Hornaday," *New York Herald Tribune*, 10 March 1937; "Claimed by Death in Stamford Home," *Stamford Advocate*, 16 January 1939. Copies of all articles are in WTHP-GT.

Helen had her parents' longevity genes and died in Stamford in November 1974 at the age of ninety-three, surviving her husband by fourteen years. George Fielding took a seat on the War Production Board during World War II while his sons served in the army. Loraine died in October 1966 at the age of fifty-five. She never wrote another book like *French Heels to Spurs*, but she did write the screenplay to *This Time for Keeps* (1947) starring Esther Williams and Jimmy Durante. She never married. In 1945, George T. Fielding, known as Dodge, was killed in action at the age of twenty-eight in the Philippines. Temple Hornaday Fielding served in the OSS during World War II and shared his grandfather's tastes for writing and travel. He authored the *Fielding Guide* series of tourism books, and died in 1983.

EPILOGUE

1. George O. Shields to WTH, 18 April 1912, box 14, WTHP-LC.
2. WTH, "The Last Call for Game Salvage," 663.
3. Allen, *Guardian of the Wild*, 30.
4. "Conservation Plan for the Eastern Pacific Stock of Northern Fur Seal (CallorhinusUrsinis)," December 2007, www.fakr.noaa.gov.
5. Aldo Leopold, "Wildlife in American Culture," 1, 2, 3, 4.

BIBLIOGRAPHY

Manuscript Collections

American Heritage Center, Laramie, Wyoming
    John Bird Burnham Papers
American Philosophical Society, Philadelphia, Pennsylvania
    Charles Davenport Papers
    Robert Cushman Murphy Papers
Denver Public Library, Denver, Colorado
    American Bison Society Papers
    Rosalie Edge Papers
Peter Edge, Winnetka, Illinois
    Rosalie Edge Papers
Hawk Mountain Sanctuary, Kempton, Pennsylvania
    Rosalie Edge Papers
Stephen Haynes, Minneapolis, Minnesota
    Chester Jackson Papers
Herbert Hoover Presidential Library, West Branch, Iowa
    Herbert Hoover Papers
Huntington Library, San Marino, California
    Letters to Albert Bigelow Paine
Indiana Historical Society, Indianapolis
    Michael H. Miller Collection
Iowa State University, Ames
    Lewis Pammel Papers
Library of Congress, Washington, D.C.
    Irving Brant Papers
    Calvin Coolidge Papers
    William T. Hornaday Papers
    Theodore Roosevelt Papers
    Woodrow Wilson Papers

New York Public Library
    National Association of Audubon Societies Papers
Plainfield Public Library, Plainfield, Indiana
    William T. Hornaday Papers
Stanford University, Stanford, California
    George A. Clark Fur Seal Controversy Papers
State Historical Society of Iowa, Des Moines
    John F. Lacey Papers
University of Arkansas, Fayetteville
    Minos Miller Papers
University of Nebraska, Lincoln
    Charles Edwin Bessey Papers
University of Rochester, Rochester, New York
    Henry A. Ward Papers
University of South Dakota, Vermillion
    Peter Norbeck Papers
University of Virginia, Charlottesville
    Samuel Marvin Papers
Wildlife Conservation Society, Bronx, New York
    American Bison Society Papers
    William T. Hornaday Papers
    Office of the Director (William T. Hornaday) Papers
    Office of the President (Madison Grant) Papers
    Office of the Secretary (Madison Grant) Papers
Yale University, New Haven, Connecticut
    George Bird Grinnell Papers

Works by William Temple Hornaday

"An African Pygmy." *New York Zoological Society Bulletin* (October 1906): 301–2.
The American Natural History: A Foundation of Useful Knowledge of the Higher Animals of North America. New York: Charles Scribner's Sons, 1904.
Awake! America. New York: Moffat, Yard, 1918.
Camp-Fires in the Canadian Rockies. New York: Charles Scribner's Sons, 1907.
Camp-Fires on Desert and Lava. New York: Charles Scribner's Sons, 1908.
"Conservation!: The Great Task for Boys." In *The Boy Scout Yearbook*, edited by Walter McGuire and Franklin K. Matthews, 36–37. New York: D. Appleton, 1915.
"The Crocodile in Florida." *American Naturalist* 9 (September 1875): 498–504.
A Democracy of Crocodiles. New York: American Guardian Society, 1919.
"The Destruction of Our Birds and Mammals: A Report on the Results of an Inquiry." *Second Annual Report of the New York Zoological Society* (1898): 77–113.
"Ethics in Hunting Game." *Forum* 57 (January 1917): 103–12.
"The Extermination of the America Bison with a Sketch of Its Discovery and Life History." In *Annual Report of the Board of Regents of the Smithsonian Institution, Showing the Operations, Expenditures, and Condition of the Institution for the Year Ending June 30, 1887*. Washington, D.C.: GPO, 1887.

"Fighting among Wild Animals." *Munsey's Magazine* 25 (April 1901): 122.
Free Rum on the Congo and What It Is Doing There. Chicago: Woman's Temperance Publication Association, 1887.
"A Great Museum Builder." *Nation* 83 (July 1905): 31–32.
Hornaday's American Natural History. New York: Charles Scribner's Sons, 1927.
"How to Observe Quadrupeds." *Christian Union* 44 (29 August 1891): 414–15.
"John F. Lacey." *Annals of Iowa* 2 (1915): 582–84.
"The Last Call for Game Salvage." In *Wildlife Restoration and Conservation: Proceedings of the North American Wildlife Conference Called by President Franklin D. Roosevelt,* 663–66. Washington, D.C.: GPO, 1936.
"The London Zoological Society and Its Garden—An Object Lesson for New York." *New York Zoological Society Annual Report* 2 (1898): 43–67.
"The Loss of a Great Opportunity." *Permanent Wild Life Protection Fund Bulletin* 1 (March 1915): 83–86.
The Lying Lure of Bolshevism. New York: American Defense Society, 1919.
The Man Who Became A Savage: A Story of Our Times. Buffalo, N.Y.: P. Paul Book Co., 1896.
"Masterpieces in American Taxidermy." *Scribner's* 72 (July 1922): 3–17.
The Minds and Manners of Wild Animals. New York: Charles Scribner's Sons, 1922.
Old-Fashioned Verses. New York: Clark and Fritts, 1919.
"On the Species of Bornean Orangs, with Notes on Their Habits." In *Proceedings of the American Association for the Advancement of Science,* edited by Frederick W. Putnam, 28:438–55. Salem: Salem Press, 1880.
Our Vanishing Wild Life: Its Extermination and Preservation. New York: New York Zoological Society, 1913.
"The Permanent Fund Establishes a Medal." *Permanent Wildlife Protection Fund Bulletin* 1 (1915): 87.
Popular Official Guide to the New York Zoological Park as Far as Completed. 6th ed. New York: New York Zoological Society, 1903.
Popular Official Guide to the New York Zoological Park. 10th ed. New York: New York Zoological Society, 1909.
"The Psychology of Wild Animals." *McClure's* 32 (February 1908): 469–79.
"Report of the Director of the New York Zoological Park to the Board of Managers." *New York Zoological Society Annual Report* 3 (1899): 47–71.
"Report of the Director of the Zoological Park." *New York Zoological Society Annual Report* 3 (1899): 39–49.
"Report of the Director of the Zoological Park." *New York Zoological Society Annual Report* 8 (1904): 50.
"Report on Character and Availability of South Bronx Park." *New York Zoological Annual Report* 1 (1897): 26–34.
"Report on the Department of Living Animals in the United States National Museum, 1888." In *Annual Report of the Board of Regents of the Smithsonian Institution, 1888.* Washington, D.C.: GPO, 1890.
"Report upon a Tour of Inspection of the Zoological Gardens of Europe." *New York Zoological Society Annual Report* 1 (1897): 35.
"The Rescued Fur Seal Industry." *Science* 52 (23 July 1920): 81–82.

"The Right Way to Teach Zoology." *Outlook* 95 (June 1910): 256–63.
A Searchlight on Germany. New York: American Defense Society, 1917.
Tales from Nature's Wonderlands. New York: Charles Scribner's Sons, 1924.
Taxidermy and Zoological Collecting. With W. J. Holland. 7th ed. New York: Charles Scribner's Sons, 1900.
Thirty Years War for Wild Life: Gains and Losses in the Thankless Task. New York: Charles Scribner's Sons, 1931.
"Twelve Years' Perspective of the Zoological Park." *New York Zoological Society Annual Report* 14 (January 1910): 99–121.
"Two Great Campaigns for Wild Life Sanctuaries." *Permanent Wild Life Protection Fund Bulletin* 3 (1 March 1920): 69–72.
Two Years in the Jungle: The Experiences of a Hunter and Naturalist in India, Ceylon, the Malay Peninsula and Borneo. New York: Charles Scribner's Sons, 1886.
"What Price Dead Ducks?" *Permanent Wild Life Protection Fund Bulletin* 12 (June 1935): 33.
"Who Is Responsible for the Exterminative Programs?" *Permanent Wild Life Protection Fund Bulletin* 3 (1920): 135–37.
"The Wichita National Bison Preserve." *New York Zoological Society Bulletin* (September 1909): 556.
Wild Animal Interviews and Wild Opinions of Us. New York: Charles Scribner's Sons, 1928.
A Wild-Animal Round-Up. New York: Charles Scribner's Sons, 1925.
Wild Life Conservation in Theory and Practice. New Haven: Yale University Press, 1914.
"Wild Life Protection." In *Proceedings from the Fourth National Conservation Congress, Indianapolis, October 1–4, Inclusive, 1912*, 344–47. Indianapolis: National Conservation Congress, 1912.

Other Sources

Adams, Charles C. *Guide to the Study of Animal Ecology.* New York: Macmillan, 1913.
Adams, Rachel. *Sideshow U.S.A.: Freaks and the American Cultural Imagination.* Chicago: University of Chicago Press, 2001.
Allen, Thomas. *Guardians of the Wild: The Story of the National Wildlife Federation, 1936–1986.* Bloomington: University of Indiana Press, 1987.
Altherr, Thomas. "The American Hunter-Naturalist and the Development of the Code of Sportsmanship." *Journal of Sports History* 5 (Spring 1978): 7–22.
Armitage, Kevin. *The Nature Study Movement: The Forgotten Popularizer of America's Conservation Ethic.* Lawrence: University Press of Kansas, 2009.
Arnold, David, and Ramachandra Guha, eds. *Nature, Culture, Imperialism: Essays on the Environmental History of South Asia.* Delhi: Oxford University Press, 1995.
Bailey, George. *Illustrated Buffalo: The Queen City of the Lakes: Past, Present, and Future.* New York: Acme Publishing and Engraving Company, 1890.

BIBLIOGRAPHY

Bailey, Thomas. "The North Pacific Sealing Convention of 1911." *Pacific Historical Review* 4 (January 1935): 1–14.

Baker, Frank. "National Zoological Park." In *The Smithsonian Institution, 1846-1896: History of Its Half Century*, edited by George Brown Goode, 443–58. Washington, D.C.: GPO, 1897.

Barney, Charles. "Report of the Executive Committee." In *Tenth Annual Report of the New York Zoological Society, 1905*, 31–43. New York: New York Zoological Society, 1906.

Barrow, Mark V. "The Specimen Dealer: Entrepreneurial Natural History in America's Gilded Age." *Journal of the History of Biology* 33 (1 December 2000): 493–534.

Baynes, Ernest J. "The Fight to Save the Buffalo." *Forest & Stream* 64 (18 February 1905): 133.

Bean, Michael J. *The Evolution of National Wildlife Law.* Rev. ed. New York: Praeger, 1983.

Bendiner, Robert. *The Fall of the Wild, The Rise of the Zoo.* New York: Dutton, 1981.

Bledstein, Burton J. *The Culture of Professionalism: The Middle Class and the Development of Higher Education in America.* New York: Norton, 1976.

Board of Regents of the Smithsonian Institution. *Annual Report, 1884.* Washington, D.C.: GPO, 1885.

———. *Annual Report, 1887.* Washington, D.C.: GPO, 1888.

———. *Annual Report, 1888.* Washington, D.C.: GPO, 1890.

———. *Annual Report, 1890.* Washington, D.C.: GPO, 1891.

———. *Annual Report, 1891.* Washington, D.C.: GPO, 1893.

Bradford, Phillips Verner, and Harvey Blume. *Ota Benga: The Pygmy in the Zoo.* New York: Dell, 1992.

Brant, Irving. *Adventures in Conservation with Franklin D. Roosevelt.* Flagstaff: Northland, 1988.

———. *Compromised Conservation: Can the Audubon Society Explain?* New York: Emergency Conservation Committee, 1930.

———. *A Last Plea for Waterfowl.* New York: Emergency Committee, 1934.

———. *Shotgun Conservation.* New York: Emergency Conservation Committee, 1931.

Bridges, William. *Gathering of Animals: An Unconventional History of the New York Zoological Society.* New York: Harper and Row, 1974.

Brinkley, Douglas. *The Wilderness Warrior: Theodore Roosevelt and the Crusade for America.* New York: Harper Collins, 2009.

Brooks, Paul. *Speaking for Nature: How Literary Naturalists from Henry Thoreau to Rachel Carson Have Shaped America.* Boston: Houghton Mifflin, 1980.

Brown, David, and Neil Carmony, eds. *Aldo Leopold's Wilderness: Selected Early Writings by the Author of "A Sand County Almanac."* Harrisburg, Pa.: Stackpole Books, 1990.

Broyles, Bill. "Adventure in the Pinacate." *Journal of Arizona History* (1987).

Bryan, William Jennings, and Mary Bryan. *The Memoirs of William Jennings Bryan.* Port Washington, N.Y.: Port Kennikat Press, 1971.

Bureau of Fisheries. *Report of the Commissioner of Fisheries for the Fiscal Year 1909 and Special Papers*. Washington, D.C.: GPO, 1911.

Burgess, Thornton W. *The Adventures of Poor Mrs. Quack*. Boston: Little, Brown, 1917.

———. *Now I Remember: The Autobiography of an Amateur Naturalist*. Boston: Little, Brown, 1960.

Burnham, John Bird. "What Are the Facts?" *Outdoor Life* 63 (January 1929): 18–19, 66.

Busch, Briton Cooper. *The War Against the Seals: A History of the North American Fishery*. Kingston: McGill-Queen University Press, 1983.

Campbell, Charles. "The Anglo-American Crisis in the Bering Sea, 1890–1891." *Mississippi Valley Historical Review* 48 (December 1961): 393–414.

———. "The Bering Sea Settlements of 1892." *Pacific Historical Review* 32 (November 1963): 347–67.

Cameron, Jenks. *The Bureau of Biological Survey: Its History, Activities, and Organization*. Baltimore: Johns Hopkins University Press, 1929.

Carnegie, Andrew. *Round the World*. New York: Charles Scribner's Sons, 1884.

Carson, Gerald. "T. R. and the 'Nature Fakers.'" *American Heritage* 22 (February 1971): 61–65, 110.

Cart, Theodore W. "The Struggle for Wildlife Protection in the United States, 1870–1900." Ph.D. diss., University of North Carolina, 1971.

Cervasco, George, and Richard P. Harmond, eds. *Modern American Environmentalists: A Biographical Encyclopedia*. Baltimore: Johns Hopkins University Press, 2009.

Cioc, Mark. *The Game of Conservation: International Treaties to Protect the World's Migratory Animals*. Athens: University of Ohio Press, 2009.

Clark, George A. "Recent Fur Seal Legislation." *Nation* 96 (3 April 1913): 333.

Colpitts, George. *Game in the Garden: A Human History of Wildlife in Western Canada to 1940*. Vancouver: University of British Columbia Press, 2002.

———. "Wildlife Promotions, Western Canadian Boosterism, and the Conservation Movement, 1890–1914." *American Review of Canadian Studies* 28 (Spring–Summer 1998): 103–30.

Cope, Edward Drinker. *Crocodilians, Lizards, and Snakes of North America*. Washington, D.C.: GPO, 1900.

Crandall, Lee. *A Zooman's Notebook*. Chicago: University of Chicago Press, 1966.

Cronon, William. *Nature's Metropolis: Chicago and the Great West*. New York: Norton, 1991.

Cutright, Paul Russell. *The Great Naturalists Explore South America*. Freeport, N.Y.: Books for Libraries Press, 1940.

———. *Theodore Roosevelt: The Making of a Conservationist*. Urbana: University of Illinois Press, 1985.

———. *Theodore Roosevelt the Naturalist*. New York. Charles Scribner's Sons, 1956.

Dall, William. *Alaska and Its Resources*. New York: Arno Press, 1870.

Danz, Harold P. *Of Bisons and Men*. Niwot: University of Colorado Press, 1997.

Department of External Affairs. *Documents on Canadian External Relations*. Vol. 1. Ottawa: Queen's Printer, 1967.

DeSormo, Maitland C. *John Bird Burnham: Klondiker, Adirondacker, and Eminent Conservationist.* Saranac Lake, N.Y.: Adirondack Yesterdays, 1978.

Ditmars, Raymond. *Thrills of a Naturalist's Quest.* New York: Macmillan, 1932.

Dolph, James Andrew. "A Dedication to the Memory of William Temple Hornaday, 1854–1937." *Journal of the Southwest* 25 (Autumn 1983): 209–12.

———. "Bringing Wildlife to the Millions: William Temple Hornaday, the Early Years: 1854–1896." Ph.D. diss., University of Massachusetts, 1975.

Dorsey, Kurkpatrick. *The Dawn of Conservation Diplomacy: United States–Canadian Wildlife Protection Treaties in the Progressive Era.* Seattle: University of Washington Press, 1998.

———. "Putting a Ceiling on Sealing: Conservation and Cooperation in the International Arena, 1909–1911." *Environmental History Review* 15 (Autumn 1991): 27–45.

Doughty, Robin. *Feather Fashions and Bird Preservation: A Study in Nature Protection.* Berkeley: University of California Press, 1975.

Dunlap, Thomas R. *Saving America's Wildlife: Ecology and the American Mind, 1850–1950.* Princeton: Princeton University Press, 1988.

Edge, Peter. "A Determined Lady." February 13, 2005. www.hawkmountain.org/default/rosalie_edge.htm.

Elliott, Henry Wood. *The History and Present Condition of the Fishery Industries: The Seal-Islands of Alaska.* Department of the Interior. Washington, D.C.: GPO, 1881.

———. "The History and Present Conditions of the Fishing Industry: The Seal Islands of Alaska." In *Tenth Census of the United States.* Washington, D.C.: GPO, 1881.

———. *Report on the Conditions of the Fur-Seal Fisheries of the Pribylov Islands in 1890.* Paris: Chamerot and Renouard, 1893.

———. *A Statement Submitted in Re the Furs Seal Herd of Alaska to the House Committee on Expenditures in the Department of Commerce, December 15, 1913.* Washington, D.C.: GPO, 1913.

Elton, Charles. *Animal Ecology and Evolution.* London: Humphrey Milford, 1930.

Fielding, Loraine Hornaday. *French Heels to Spurs.* New York: Century, 1930.

Fischer, David Hackett. *Albion's Seed: Four British Folkways in America.* New York: Oxford University Press, 1989.

Flader, Susan. *Thinking Like a Mountain: Aldo Leopold and the Evolution of Our Ecological Attitude toward Deer, Wolves, and Forestry.* Columbia: University of Missouri Press, 1974.

Flores, Dan. "Bison Ecology and Bison Diplomacy: The Southern Plains from 1800 to 1850." *Journal of American History* 78 (September 1991): 465–85.

Foord, John. *The Life and Public Services of Andrew Haswell Green.* Garden City, N.Y.: Doubleday, Page, 1913.

Forbes, John Ripley. *In the Steps of the Great American Zoologist: William Temple Hornaday.* New York: M. Evans, 1966.

Forbush, Edward Howe. *A History of the Wild Game Birds, Wild-Fowl, and Shore Birds of Massachusetts and Adjacent States.* Boston: Wright and Potter, 1912.

Fox, Stephen. *John Muir and His Legacy: The American Conservation Movement.* Boston: Little, Brown, 1981.
Furmansky, Dyana Z. *Rosalie Edge, Hawk of Mercy: The Activist Who Saved Nature from the Conservationists.* Athens: University of Georgia Press, 2009.
Gabrielson, Ira Noel. *Wildlife Refuges.* New York: Macmillan, 1943.
Gard, Wayne. *The Great Buffalo Hunt.* New York: Knopf, 1959.
Gaustad, Edwin S., ed. *The Rise of Adventism: Religion and Society in Mid-Nineteenth Century America.* New York: Harper and Row, 1974.
Gifford, John Garry. *The Citizen Soldiers: The Plattsburgh Training Camp Movement, 1913–1920.* Louisville: University of Kentucky Press, 1972.
Graham, Frank. *The Audubon Ark: A History of the National Audubon Society.* New York: Knopf, 1990.
——. *Man's Dominion: The Story of Conservation in America.* Philadelphia: Lippincott, 1971.
Grant, Madison. "A Brief History of the Landscape and Forestry Work in the New York Zoological Park." *New York Zoological Annual Report* 9 (1905): 41–46.
——. *The Passing of the Great Race.* 1916. Reprint, New York: Arno Press, 1970.
Grinnell, George Bird. "A Plank." *Forest & Stream* 42 (3 February 1894): 89.
Grinnell, George Bird, and Theodore Roosevelt, eds. *American Big-Game Hunting: The Book of the Boone and Crockett Club.* New York: Forest & Stream Publishing, 1893.
Grinnell, George Bird, and Charles Sheldon, eds. *Hunting and Conservation: The Book of the Boone and Crockett Club.* 1925. Reprint, New York: Arno Press, 1970.
Grinnell, Josiah Bushnell. *Events of Forty Years: Autobiographical Reminiscences of an Active Career from 1850 to 1890.* Boston: D. Lothrop, 1891.
Guggisberg, C. A. W. *Crocodiles: Their Natural History, Folklore, and Conservation.* Harrisburg, Pa.: Stackpole, 1972.
Guterl, Matthew Pratt. *The Color of Race in America, 1900–1940.* Cambridge: Harvard University Press, 2001.
Ham, Ken, and Charles A. Ware. *Darwin's Plantation: Evolution's Racist Roots.* Green Forest, Ark.: Master Books, 2007.
Hancocks, David. *A Different Nature: The Paradoxical World of Zoos and their Uncertain Future.* Berkeley and Los Angeles: University of California Press, 2001.
Harper, Frank. "John McFarlane Phillips." In *Pittsburgh of Today: Its Resources and People,* vol. 4. New York: American Historical Society, 1931.
Hayden, Sherman Strong. *The International Protection of Wildlife: An Examination of Treaties and Other Agreements for the Preservation of Birds and Mammals.* New York: Columbia University Press, 1942.
Hays, Samuel P. *Beauty, Health, and Permanence: Environmental Politics in the United States, 1955–1985.* Cambridge: Cambridge University Press, 1987.
——. *Conservation and the Gospel of Efficiency: The Progressive Conservation Movement, 1890–1920.* Cambridge: Harvard University Press, 1968.
Herman, Daniel Justin. "The Hunter's Aim: The Cultural Politics of American

Sport Hunters, 1880–1910." *Journal of Leisure Research* 35, no. 4 (Fourth Quarter 2003): 455–75.

———. *Hunting and the American Imagination.* Washington, D.C.: Smithsonian Institution Press, 2001.

Hilkey, Judy. *Character Is Capital: Success Manuals and Manhood in Gilded Age America.* Chapel Hill: University of North Carolina Press, 1997.

History of Hendricks County, Indiana. Chicago: Inter-State Publishing, 1885.

Hoage, R. J., and William Deiss, eds. *New Worlds, New Animals: From Menageries to Zoological Parks in the Nineteenth Century.* Baltimore: Johns Hopkins University Press, 1996.

Holmes, S. J. *The Evolution of Animal Intelligence.* New York: Henry Holt, 1911.

Hoover, Herbert. "Conservation of Our Waterfowl, August 25, 1931." In *Herbert Hoover: Proclamations and Executive Orders*, 1:156–57. Washington, D.C.: GPO, 1974.

———. "Statement Announcing the Proposal of the Moratorium on Intergovernmental Debts, June 21, 1931." In *Public Papers of the Presidents of the United States, Herbert Hoover 1931*, 325–27. Washington, D.C.: GPO, 1976.

Hornaday, Aline, and Joyce Poe. *The Hornadays, Root and Branch: The Sequel.* N.d. http://members.cox.net/annelwood/.

Hornaday, Quinn. *The Hornadays Root and Branch.* Los Angeles: Stockton Trade Press, 1979.

Horowitz, Helen Lefkowitz. "Animal and Man in the New York Zoological Park." *New York History* 55 (1975): 426–55.

———. "The National Zoological Park: 'City of Refuge' or Zoo?" *Records of the Columbia Historical Society* 49 (1973/74): 405–29.

Illustrated Buffalo: The Queen City of the Lakes: Past, Present, and Future. New York: Acme Publishing and Engraving Co., 1890.

Irmscher, Christoph. *The Poetics of Natural History: From John Bartram to William James.* New Brunswick, N.J.: Rutgers University Press, 1999.

Isenberg, Andrew C. *The Destruction of the Bison: An Environmental History, 1750–1920.* New York: Cambridge University Press, 2000

———. "The Returns of the Bison: Nostalgia, Profit, and Preservation." *Environmental History* 2 (April 1997): 179–96.

Jacoby, Karl. *Crimes Against Nature: Squatters, Poachers, Thieves, and the Hidden History of Conservation.* Berkeley and Los Angeles: University of California Press, 2001.

James, Lawrence. *Raj: The Making and Unmaking of British India.* New York: St. Martin's Press, 1997.

Jardine, William. "The Year in Agriculture: The Secretary's Report to the President." In *Agriculture Yearbook, 1925.* Washington, D.C.: GPO, 1926.

Jenkins, Alan C. *The Naturalists: Pioneers of Natural History.* New York: Mayflower Books, 1978.

Jordan, David Starr. *The Days of a Man: Being Memories of a Naturalist, Teacher, and Minor Prophet of Democracy.* Yonkers-on-Hudson: World Book, 1922.

Jordan, David Starr, Vernon Lyman Kelly, and Harold Heath. *Animal Studies: A*

Textbook for Elemental Zoology in High Schools and Colleges. New York: D. Appleton, 1903.

Kimball, David, and Jim Kimball. *The Market Hunter.* Minneapolis: Dillon Press, 1969.

Kisling, Vernon N. *Zoo and Aquarium History: Ancient Animal Collections to Zoological Gardens.* Boca Raton, Fla.: CRC Press, 2001.

Land, Gary, ed. *Adventism in America: A History.* Grand Rapids, Mich.: Eerdmans, 1986.

Lears, T. Jackson. *No Place of Grace: Anti-Modernism and the Transformation of American Culture, 1880–1920.* Chicago: University of Chicago Press, 1981.

Lendt, David L. *Ding: The Life of Jay Norwood Darling.* Ames: Iowa State University Press, 1979.

Leopold, Aldo. "Wildlife in American Culture." *Journal of Wildlife Management* 7 (January 1943): 1–16.

Loo, Tina. "Of Moose and Men: Hunting for Masculinity in British Columbia, 1880–1939." *Western Historical Quarterly* 32 (Autumn 2001): 297–319.

Loring, J. Alden. "The Wichita Buffalo Range." *New York Zoological Society Annual Report* 10 (1906): 181–200.

Lucas, Frederic A. *Fifty Years of Museum Work: Autobiography, Unpublished Papers, and Bibliography.* New York: American Museum of Natural History, 1933.

———. "The Mounting of Mungo." *Science* 8 (15 October 1886): 337–41.

Lutts, Ralph H. *The Nature Fakers: Wildlife Science and Sentiment.* New York: Fulcrum, 1990.

MacKenzie, John. *Empire of Nature: Hunting, Conservation, and British Imperialism.* Manchester: University of Manchester Press, 1997.

Macleod, David. *Building Character in the American Boy: The Boy Scouts, YMCA, and Their Forerunners, 1870–1920.* Madison: University of Wisconsin Press, 2004.

Madden, Dave. *The Authentic Animal: Inside the Odd and Obsessive World of Taxidermy.* New York: Macmillan, 2011.

Malthus, Thomas R. *An Essay on the Principle of Population and a Summary View of the Principle of Population.* 1798. Reprint, edited by Anthony Flew, Baltimore: Penguin, 1976.

Matthiessen, Peter. *Wildlife in America.* New York: Viking Penguin, 1987.

McHugh, Tom. *The Time of the Buffalo.* New York: Knopf, 1972.

McIver, Stuart. *Death in the Everglades: The Murder of Guy Bradley, America's First Martyr to Environmentalism.* Gainesville: University Press of Florida, 2003.

Meine, Curt. *Aldo Leopold: His Life and Work.* Madison: University of Wisconsin Press, 1988.

Mencken, H. L. *The Letters of H. L. Mencken.* Edited by Guy Forgue. Boston: Northeastern University Press, 1981.

Merkel, Herman. "The New York Idea of a Zoological Park." *American City* 9 (October 1913): 298–302.

Mighetto, Lisa. *Wild Animals and American Environmental Ethics.* Tucson: University of Arizona Press, 1991.

Miller, Char. *Gifford Pinchot and the Making of Modern Environmentalism.* Washington, D.C.: Island Press, 2001.

Morgan, Douglas. *Adventism and the American Republic: The Public Involvement of a Major Apocalyptic Movement.* Knoxville: University of Tennessee Press, 2001.

Murray, Robert K. *The Red Scare: A Study in National Hysteria, 1919–1920.* Minneapolis: University of Minnesota Press, 1955.

Nash, Roderick. *Wilderness and the American Mind.* 3rd ed. New Haven: Yale University Press, 1982.

Newton, Julianne Lutz. *Aldo Leopold's Odyssey.* Washington, D.C.: Island Press, 1982.

Nixon, Edgar B., ed. *Franklin D. Roosevelt and Conservation, 1911–1945.* 2 vols. New Hyde Park, N.Y.: General Services Administration National Archives and Research Services, 1952.

Nostrand, Jeanne Van. "The Seals Are About Gone. . . ." *American Heritage* 14 (June 1963): 11–16, 78–80.

Orr, Oliver H., Jr. *Saving American Birds: T. Gilbert Pearson and the Founding of the Audubon Movement.* Gainesville: University of Florida Press, 1992.

Osborn, Henry Fairfield. *The American Museum of Natural History: Its Origin, Its History, the Growth of Its Departments to December 31, 1909.* New York: Irving Press, 1911.

———. "Preservation of the Wild Animals of North America." In *Boone and Crockett Club Annual Meeting.* Washington, D.C., 1904.

———. "Progress of the New York Zoological Park." *Science* 11 (22 June 1900): 963–65.

Overfield, Richard. *Science with Practice: Charles E. Bessey and the Maturing of American Botany.* Ames: Iowa State University Press, 1993.

Palmer, T. S. *Legislation for the Protection of Birds Other Than Game Birds.* Washington, D.C.: GPO, 1900.

———. "Report of the Acting Chief of the Division of Biological Survey." In *Annual Reports of the Department of Agriculture for the Fiscal Year Ended June 30, 1900,* 35–48. Washington, D.C.: GPO, 1900.

The Parliamentary Debates (House of Lords). Vol. 12. 5th series. London: His Majesty's Stationery Office, 1913.

Paterson, D. G. "The North Pacific Seal Hunt, 1886–1910: Rights and Regulations." *Explorations in Economic History* 14 (April 1977): 97–114.

Pearlman, Michael. *To Make Democracy Safe for America: Patricians and Preparedness in the Progressive Era.* Urbana: University of Illinois Press, 1984.

Pearson, Thomas Gilbert. *Adventures in Bird Protection.* New York: D. Appleton-Century, 1937.

Peterson, John M., ed. "Buffalo Hunting in Montana in 1886: The Diary of W. Harvey Brown." *Montana: The Magazine of Western History* 31 (Autumn 1981): 2–13.

Pinchot, Gifford. *Breaking New Ground*. New York: Harcourt, Brace, 1947.

———. *The Fight for Conservation*. 1910. Reprint, Seattle: University of Washington Press, 1967.

Punke, Michael. *Last Stand: George Bird Grinnell, the Battle to Save the Buffalo, and the Birth of the New West*. New York: Harpers Collins, 2007.

Rangarajan, Mahesh. *Fencing the Forest: Conservation and Ecological Change in India's Central Provinces, 1860–1914*. Delhi: Oxford University Press, 1996.

Regal, Brian. *Henry Fairfield Osborn: Race, and the Search for the Origins of Man*. Aldershot, Hants, U.K.: Ashgate, 2002.

Reiger, John F. *American Sportsmen and the Origins of Conservation*. New York: Winchester Press, 1975.

Richardson, James D. *A Compilation of the Messages and Papers of the Presidents*. New York: Bureau of National Literature, 1911.

———. *A Compilation of the Messages and Papers of the Presidents*. New York: Bureau of National Literature, 1927.

Roe, Frank Gilbert. *The North American Buffalo: A Critical Study of the Species in Its Wild State*. 2nd ed. Toronto: University of Toronto Press, 1970.

Roosevelt, Theodore. *An Autobiography*. 1913. Reprint, New York: Da Capo Press, 1985.

———. *Hunting Trips of a Ranchman: Sketches of Sport on the Northern Cattle Plains*. G. P. Putnam's Sons, 1885.

———. *The Letters of Theodore Roosevelt*. Edited by Elting E. Morison. 8 vols. Cambridge: Harvard University Press, 1951–54.

———. *Literary Essays*. New York: Charles Scribner's Sons, 1927.

Roppel, Alton Y., and Stuart P. Davey. "Evolution of Fur Seal Management on the Pribilof Islands." *Journal of Wildlife Management* 29 (1965): 448–63.

Rosenberg, Morton. *Iowa on the Eve of the Civil War: A Decade of Frontier Politics*. Norman: University of Oklahoma Press, 1972.

Runte, Alfred. *National Parks: The American Experience*. 2nd rev. ed. Lincoln: University of Nebraska Press, 1987.

Sage, Leland. *A History of Iowa*. Ames: Iowa State University Press, 1974.

Sanborn, Elwin R. "An Object Lesson in Bison Preservation." *Zoological Society Bulletin* 16 (May 1913): 990–93.

Scarpino, Philip. *Great River: An Environmental History of the Upper Mississippi, 1890–1950*. Columbia: University of Missouri Press, 1985.

Schmitt, Peter. *Back to Nature: The Arcadian Myth in Urban America*. New York: Oxford University Press, 1969.

Schorger, A. W. *The Passenger Pigeon: Its Natural History and Extinction*. Norman: University of Oklahoma Press, 1973.

Sellers, Charles C. *Charles Willson Peale*. New York: Charles Scribner's Sons, 1969.

———. *Mr. Peale's Museum: Charles Willson Peale and the First Popular Museum of Natural Science and Art*. New York: Norton, 1980.

Shell, Hanna Rose. "Introduction: Finding the Soul in the Skin." Introduction to William Temple Hornaday, *The Extermination of the American Bison*, edited by E. Anne Bolen. Washington, D.C.: Smithsonian Institution Press, 2002.

Shelton, Brenda Kurtz. *Reformers in Search of Yesterday: Buffalo in the 1890s.* Albany: State University of New York Press, 1976.
Sherwood, Morgan. *Big Game in Alaska: A History of Wildlife and People.* New Haven: Yale University Press, 1981.
———. "Seal Poaching in the North Pacific." *Alaska History* 1 (1984): 44–51.
Shields, George O. "Hunting Caribou with a Camera." *Shields' Magazine* 2 (May 1906): 305–6.
Shipley, Donald DeVries. "A Study of Conservation Philosophies and Contributions of Some Important American Conservation Leaders." Ph.D. diss., Cornell University, 1953.
Sobel, Robert. *Coolidge: An American Enigma.* Washington, D.C.: Regnery, 1998.
Society of American Taxidermists. *Third Annual Report of the Society of American Taxidermists.* Washington, D.C.: Gibson Brothers, Printers, 1884.
Spence, Benjamin A. "The National Career of John Wingate Weeks." Ph.D. diss., University of Wisconsin, 1971.
Spiro, Jonathan Peter. *Defending the Master Race: Conservation, Eugenics, and the Legacy of Madison Grant.* Hanover, N.H.: University Press of New England, 2009.
Stampp, Kenneth. *Indiana Politics during the Civil War.* Indianapolis: Indiana Historical Society, 1949.
State of New York. *Public Papers of John A. Dix, Governor, 1911.* Albany: J. B. Lyon Co., 1912.
The Statutes at Large of the United States of America from March 1911 to March 1913. Washington, D.C.: GPO, 1913.
Sterling, Keir. *Last of the Naturalists: The Career of C. Hart Merriam.* New York: Arno Press, 1977.
Stott, Jeffrey. "An American Idea of a Zoological Park: An Intellectual History." Ph.D. diss., University of California at Santa Barbara, 1981.
Swain, Donald. *Federal Conservation Policy, 1921–1933.* Berkeley: University of California Press, 1963.
Sykes, Godfrey G., and Diane Boyer. "The 'Lost' 1907 Pinacate Diary of Godfrey G. Sykes." *Journal of the Southwest* 49 (July 2007): 165–87.
Thomas, William H. *Unsafe for Democracy: World War I and the U.S. Justice Department's Covert Campaign to Suppress Dissent.* Madison: University of Wisconsin Press, 2008.
Tober, James. *Who Owns the Wildlife? The Political Economy of Conservation in Nineteenth-Century America.* Westport, Ct.: Greenwood Press, 1981.
Townsend, C. H. "The Pribilof Fur Seal Herd and the Prospects for Its Increase." *Science* 34 (27 October 1911): 568–70.
Trefethen, James. *An American Crusade for Wildlife.* New York: Winchester Press, 1975.
Turner, James. *Reckoning with the Beast: Animals, Pain, and Humanity in the Victorian Mind.* Baltimore: Johns Hopkins University Press, 1980.
Turner, Raymond M. "Long-Term Vegetation Change at a Fully Protected Sonoran Desert Site." *Ecology* 71 (April 1990): 464–77.

U.S. Congress (House). *Seal Islands of Alaska.* Vol. 5. Reports. Washington, D.C.: GPO, 1911.

U.S. Congress (Senate). *Alaskan Seal Fisheries, Hearings before the Committee on Conservation of Natural Resources, United States Senate on Bill S.9959, February 4, 1911.* Washington, D.C.: GPO, 1911.

U.S. House Committee on Ways and Means, and Oscar Wilder Underwood. *Tariff Schedules: Hearings before the Committee on Ways and Means, House of Representatives; Vol. V, Schedules M and N. Index in Final Volume. January 17, 1913.* Washington, D.C.: GPO, 1913.

U.S. Senate Committee on Finance. *Tariff Schedules: Hearings before the Subcommittees of the Committee on Finance, United States Senate, Sixty-third Congress, First Session on H.R. 3321, an Act to Reduce Tariff Duties and to Provide Revenue for the Government, and for Other Purposes.* Washington, D.C.: GPO, 1913.

———. *The Fur Seals and Other Life of the Pribiloff Islands, Alaska in 1914.* Vol. 6. Documents. Washington, D.C.: GPO, 1915.

Van Name, Willard G., Dewitt Miller, and Davis Quinn. *A Crisis in Conservation.* 4th ed. New York: Emergency Conservation Committee, 1931.

Verner, Samuel P. "The Story of Ota Benga." *New York Zoological Society Bulletin* 19 (19 July 1916): 1377–79.

Vileisis, Ann. *Discovery of the Unknown Landscape: A History of America's Wetlands.* Washington, D.C.: Island Press, 1997.

Ward, Roswell Howell. *Henry A. Ward, Museum Builder to America.* Rochester, N.Y.: Rochester Historical Society, 1948.

Warren, Louis. *The Hunter's Game: Poachers and Conservationists in Twentieth-Century America.* New Haven: Yale University Press, 1997.

Washington, Harriet A. *Medical Apartheid: The Dark History of Medical Experimentation on Black Americans From Colonial Times to the Present.* New York: Harlem Moon, 2006.

Weibe, Robert. *The Search for Order, 1877–1920.* New York: Hill and Wang, 1967.

Welker, Robert Henry. *Birds and Men: American Birds in Science, Art, Literature, and Conservation, 1800–1900.* Cambridge: Belknap Press of Harvard University Press, 1955.

———. *Natural Man: The Life of William Beebe.* Bloomington: Indiana University Press, 1975.

Wilcove, David S. *The Condor's Shadow: The Loss and Recovery of Wildlife in America.* New York: W. H. Freeman, 1999.

Wild, Peter, ed. *The Opal Desert: Explorations of Fantasy and Reality in the American Southwest.* Austin: University of Texas Press, 1999.

Williams, Gerald W. *The Forest Service: Fighting for Public Lands.* Westport, Ct.: Greenwood, 2007.

Wilson, Woodrow. *The Papers of Woodrow Wilson.* Edited by Arthur S. Link. 68 vols. Vols. 27, 28, 47, and 51. Princeton: Princeton University Press, 1966–94.

Worster, Donald. *Nature's Economy: A History of Ecological Ideas.* New York: Cambridge University Press, 1985.

———. *A Passion for Nature: The Life of John Muir.* New York: Oxford University Press, 2008.

Yearbook of the United States Department of Agriculture, 1900. Washington, D.C.: GPO, 1901.

Yearbook of the United States Department of Agriculture, 1906. Washington, D.C.: GPO, 1907.

Yearbook of the United States Department of Agriculture, 1908. Washington, D.C.: GPO, 1909.

Yearbook of the United State Department of Agriculture, 1910. Washington, D.C.: GPO, 1911.

Yearbook of the United States Department of Agriculture, 1911. Washington, D.C.: GPO, 1912.

Yearbook of the United State Department of Agriculture, 1913. Washington, D.C.: GPO, 1914.

Yearbook of the United States Department of Agriculture, 1914. Washington, D.C.: GPO, 1915.

Zontek, Ken. *Buffalo Nation: American Indian Efforts to Restore the Bison.* Lincoln: University of Nebraska Press, 2007.

INDEX

Adventism, 14–16. *See also* Hornaday, William Temple (WTH): influence: of Adventism on
Adventures in Bird Protection (Pearson), 119, 169
Adventures of Poor Mrs. Quack, The (Burgess), 146
Alaska: Conservation Act, 6; Game Act, 155; game regulations, 159–61, 164; Sea Mammal Protection Act, 6; Sulzer bill, 154–56; *See also* fur seals and their conservation
Allen, Joel, 65, 118, 120
American Bisons, The (Allen), 65
American Defense Society, 142, 151–52
American Game Protection Association (AGPA), 120–22, 147, 161, 165, 169, 171–72, 177, 179–81, 200
American Guardian Society, 151
American Museum of Natural History, 76, 101, 182
American Natural History, The (WTH), 20, 71, 87–91, 98, 107, 137
American Naturalist, 26, 90–91
American Ornithological Union, 119
American Wildlife Conference, 196, 199–200
Animal Intelligence (Romanes), 162
Anthony, Daniel, 161
Antiquities Act, 3, 92
Army League, 142
Aubrey, Charles, 105

Audubon, John J., 21, 71
Audubon Society. *See* National Association of Audubon Societies
Auten, Ben, 18
Autobiography, An (T. Roosevelt), 92–93
Avery, Carlos, 179, 184
Awake America! (WTH), 151

Baird, Spencer, 60, 63, 67
Bayne, Howard K., 111–13
Baynes, Ernest Harold, 104
Beard, Daniel Carter, 57, 198
Beck Committee, 194–95
Beebe, William, 80, 86
Beecher, Henry Ward, 27
Benga, Ota, 96–98
Bessey, Charles E., 19–22, 23, 56, 83
Birds of North America (Audubon), 21
Blair, W. Reid, 86, 173–75, 195, 197–98
Blume, Harvey, 97
Boone and Crockett Club, 76, 92, 114, 140, 153–54, 188; WTH's membership in, 154
Bowers, George, 108
Boy Scouts, 57, 157, 198
Bradford, Phillps Verner, 97
Bradley, Guy, 131
Brant, Irving, 180, 183–86, 191, 196, 201; view of WTH, 187–88
Bronx Zoo. *See* New York Zoological Park (Bronx Zoo)

INDEX

Brown, William Harvey, 62, 64
Bryan, William Jennings, 16
buffalo (bison) and their conservation, 1, 4–5, 7, 66; American Bison Society, 104–6; Flathead Bison Preserve, 104–6, 164; role of Native Americans, 106; saved from extinction, 106; Wichita Bison Preserve, 91–94, 104, 200
Bureau of Biological Survey (U.S.), 6, 16, 84, 112, 124, 126, 141, 153, 155, 164, 166–68, 174, 176, 188, 196
Burgess, Thornton, 146, 156–57, 181, 201
Burnham, Henry, 129
Burnham, John Bird, 102, 120–22, 125–26, 128–30, 159, 161, 163–65, 167–73, 179–80, 182, 186, 191; asks Madison Grant to restrain WTH, 171; attacks WTH's masculinity, 164; sues WTH, 177–78
Burroughs, John, 2, 88, 131, 162
Bynum, Dr., 101

California Audubon Society, 143
Call to Action, A (WTH), 190
Camp Fire Club, 3
Camp-Fires in the Canadian Rockies (WTH), 98–99, 103, 156
Camp-Fires on Desert and Lava (WTH), 103–4, 156
Cannon, Joe, 68–69
"Canoe and Rifle on the Orinoco" (WTH), 59
Careless Chicken, The, 17
Carmody, Thomas, 136, 137
Carnegie, Andrew, 48–49, 50, 56–57, 79, 102, 119; capitalizes New York Zoological Society pension fund, 87; finances photographs for *American Natural History*, 89; purchases WTH books for libraries, 91; subscription to *Our Vanishing Wildlife*, 127
Chamberlain, George, 144
Chapman, Frank, 118, 130–31, 162, 188

Charles Scribner's Sons, 60, 88–89, 91, 98, 137, 175, 192
Clapp, Moses, 133
Clark, George A., 123–25, 135
Colored Orphan Asylum (Brooklyn, N.Y.), 98
"Compromised Conservation" (Brant), 185
Conservation and the Gospel of Efficiency (Hays), 129
Coolidge, Calvin, 181, 189
Cope, Edward Drinker, 26
Copeland, Royal, 171
Crandall, Lee, 81, 86, 195, 198
"Crisis in Conservation, A" (Miller, Quinn, Van Name), 182–84
Custer, George Armstrong, 63
Cuvier, Baron Georges, 26

Daniels, Josephus, 148
Darling, Jay "Ding," 175, 194–96, 200, 201
Darwin, Charles, 24, 27, 202
Davenport, Charles, 105, 176
Davison, Charles, 154
Department of Agriculture (U.S.), 6, 141, 145, 166–68, 177, 189–90, 194
Destruction of Our Birds and Mammals, The (WTH), 83–85, 127
Dilg, Will, 165–69
Ditmars, Raymond, 86, 195, 198
Dix, John, 113
Dixon, Joseph, 104, 107–8
Dolph, James Andrew, 39, 65, 70
Donaldson, Thomas, 68
Du Chaillu, Paul, 23–24, 52
Ducks Unlimited, 191
Dunlap, R. W., 190
Dutcher, William, 117, 126

Eastman, George, 119
Edge, Peter, 184, 187
Edge, Rosalie, 182–88, 196, 201; view of WTH, 187–88
"Eighty Fascinating Years" (WTH), 11, 57, 64, 69, 85, 148, 176, 192–94

INDEX

Elk River Game Preserve, 99–101, 106
Elliott, Henry Wood, 106–11, 123–24, 144, 149, 154, 201
Elrod, Morton, 104–5
Emergency Conservation Committee, 184–85, 187–88, 191, 196
Endangered Species Act, 6, 202
"Extermination of the American Bison, The" (WTH), 65–66, 85, 116, 127

Fernie Game Protection Association, 99–100
Fielding, Dodge, 194, 196, 227n42
Fielding, George, 101, 149, 158, 192, 226n42
Fielding, Lorraine, 194, 226n42
Fielding, Temple, 153, 194, 196, 227n42
Forbes, John Ripley, 176
Forbush, Edward Howe, 164, 172
Forest & Stream, 72, 76, 90, 110, 120, 147
Fourth National Conservation Congress, 126
French Acclimatization Society, 136
French Heels to Spurs (Fielding), 194
Frick, Henry Clay, 114, 119
Fur Seal Advisory Board, 3, 6, 107–10, 123, 126
fur seals and their conservation, 3, 107–10, 122–25, 201; Fur Seal Convention, 122; Marine Mammal Protection Act, 201; pelagic sealing, 2, 3, 107, 109, 122–23
Fur Seal Service, 109–10, 144

game sanctuary plan, 139–41, 142–48
Gates, Moody, 157
Getting on in the World (Mathews), 58
Gilded Age, The (Twain and Warner), 131
Goode, George Brown, 53, 63, 66–68
Graham, Frank, 113
Grant, Madison, 75–80, 83, 85–86, 91, 95, 96–98, 110, 112, 114, 119, 122, 133, 138, 154–55, 160, 167, 170–72, 174–75, 188, 198; background, 75; embarrassed by WTH, 102; racial views and conservation, 7, 75–76; restrains WTH, 112, 118
Graves, Henry, 141, 144–45
Great Gatsby, The (Fitzgerald), 75
Green, Andrew H., 79
Grinnell, George Bird, 66, 97, 102, 105–6, 118–22, 128, 130, 135, 159–60, 165, 170–71, 174; criticizes WTH to Grant, 119, 160; relationship with WTH, 119, 159–60; tells Grant and Osborn that WTH is all wrong on shotguns, 121

Hagenbeck, Carl, 4, 78, 82–83
Haugen, Gilbert, 184
Hayden, Carl, 145
Hays, Samuel, 129, 140
Herrick, Myron, 136
Hetch-Hetchy Valley, 3, 107
Hirschfield, David, 86, 158–59
Hoover, Herbert, 175, 184, 189–92
Hornaday, Calvin, 10, 18, 19, 72, 96
Hornaday, Clark, 10, 12, 13
Hornaday, Helen Ross (Fielding), 54, 73, 101, 181, 192, 226n42
Hornaday, Josephine Chamberlain, 33, 37, 39, 45, 48, 64, 75, 96, 122, 141, 148, 150, 163, 181, 198; marries WTH, 53–54; meets WTH, 31; supports WTH's resignation from National Zoological Park (U.S.), 70
Hornaday, Martha Varner Miller, 10–11, 13–14, 15, 17–18
Hornaday, Mary, 10, 14, 54, 153, 158
Hornaday, Silas, 10, 11
Hornaday, William, Sr., 10, 11, 12, 13–15, 18
Hornaday, William Temple (WTH): alleges American Game and Protection Association and John Burnham are tools of the gun makers, 121–22, 125, 165, 167–68, 173, 186; alleges millinery industry controls United States Senate,

Hornaday, William Temple (*cont.*) 134; alleges Edwin Nelson was under external influence, 177; anthropomorphizes wildlife, 90, 93, 162–63, 175, 202; attacks National Association of Audubon Societies for accepting gunmaker's subscription money, 118–20; attacks Charles Sheldon, 154; attitude toward Rothermel Committee, 110; asked to write plumage ban, 132; baits Secretary of Commerce and Labor Charles Nagel, 109; birth of, 10; challenges local conservationists in British Columbia, 99–100; childhood of, 10–15, 16–18; college education of, 19–22, 70; compared to John Muir, 5, 107; complains to Secretary of Agriculture Houston about Bureau of Biological Survey policies, 155; compromises with game meat dealers in Bayne bill, 112–13; concern over Americans becoming divorced from nature, 87, 98; conducts inventory of buffalo skins in 1886, 60–61; as conservation lobbyist, 5, 106, 114–15, 134, 165; convenes meeting of conservationists at Century Club to discuss migratory bird protection, 125–26; criticizes John Burnham's management of the migratory bird bill, 125–26; criticizes "joke" bag limits, 159–60; critiques consumption, 7; critiques youth culture, 193–94; curses Reed Smoot, 146; death and funeral of, 197; death of parents of, 18; declines invitation to join the ECC, 184; description of, 21; "discovers" new crocodile species, 26; divides wildlife protection movement in dispute with Burnham, 170–71; "Eastern Establishment," description of, 164, 167, 168–69; effect of Theodore Roosevelt's death on, 156; effect of World War I on views of humanity of, 193; encourages Brant to not attack Madison Grant or the Boone and Crockett Club, 188; enters wildlife conservation, 64; exploration of the Sonora Desert by, 101–3; expresses gratitude towards Ward, 57; fears Walcott Committee trying to undo Hoover proclamation, 190–91; fights with Dr. Bynum over antlers, 101–2; fights with Minos Miller over Civil War letters, 13; fights with Oldys over credit for plumage ban, 135–36; "finds himself" at Ames, 22, 175; forms American Bison Society, 104; as grandfather, 153, 194; hires Irving Brant as lobbyist for Permanent Wildlife Protection Fund, 181; incorrect on provisions of Alaska Game Act, 155; introduces alternative to Burnham's migratory bird refuge bill, 166; introduces bag limit reductions at Migratory Bird Advisory Board, 163; joins Edge at Audubon annual meeting, 185–86; lectures at Yale, 137, 139; locates misplaced Migratory Bird Treaty, 145–46; marriage to Josephine Chamberlain Hornaday, 53–54; meets with Henry Graves on the game sanctuary plan, 141, 145; meets with H. S. Leonard of Winchester, 116–17; meets with representatives of the millinery industry, 132; mocks George Bird Grinnell, 110; and money, 18, 58–59, 153; and National Zoological Park (U.S.), 66–70; negotiates with Michael Pablo for purchase of buffalo, 105–6; nervous as public speaker, 64, 133, 137; as New York Zoological Park director, 4, 79, 74–75, 77–79, 80–82, 85–88, 90, 150–51, 158; opposes Duck Stamp Tax act, 195, 200;

opposes public shooting grounds, 165, 169, 179, 181, 182; personalizes Alaska game regulations, 160; personalizes National Association of Audubon Societies subscription scandal, 120; personalizes seal conservation, 3, 109, 124–25; praises President Hoover's conservation proclamation, 190; praises President Hoover's debt moratorium, 189; preparedness activities of, 141–42, 148–50; proposes Africa expedition, 23–24; proposes closed season on land sealing, 107–8; proposes national zoological park, 67; proposes tax on shotgun shells, 137–39, 191; publishes letters from Secretary of Commerce and Labor Charles Nagel, 109; reacts to failure of the game sanctuary plan, 146–47; reacts to Henry Fairfield Osborn's suggestion to follow Hagenbeck reforms, 82–83; real estate business in Buffalo, N.Y., 72; receives honorary doctorate, 96; Red Scare activities of, 152–53, 158; represents non-hunters in conservation, 4, 5, 165, 170, 173; responds to George Clarke's attack, 124–25; responds to Henry Wood Elliott's appeal to save seals, 107; responds to George Pratt's letter to Rosalie Edge, 187; retirement as New York Zoological Park director, 172–75; role in founding of Society of American Taxidermists, 54; role in Ota Benga controversy, 96–98; service on Migratory Bird Advisory Board, 130, 163; and sportsmen, 147–48; strategy to protect fur seals, 108; strategy to secure passage of migratory bird protection, 126; sued by Burnham, 177–78; support of women's suffrage, 54, 73; supports Alaska game revision after Sulzer bill, 155–56; supports Emergency Conservation Committee, 187; tours the west for game sanctuary plan, 142–43; vows never to fight under banner of New York Zoological Society, 115; warns Norbeck about the flaws of his bill, 179; writes game sanctuary plan bill to avoid mistakes of the past, 140

—collecting expeditions: to Asia, 31–51; —, financial troubles during, 35, 42, 44, 45, 48; —, frustrations with Ward during, 33, 42, 43–44; —, returns to the United States from, 49–50; to Caribbean and Orinoco River Valley, 27–30, 206n48; to Ceylon, 43–45; to Cuba and Florida, 24–25; to Europe with Henry Ward, 33–34; to India, 35–43; to Malaysia and Singapore, 45–46, 48; to North Africa with Henry Ward, 34

—hunting: in British Columbia, 95–96; buffalo, 61–63, 106; elephants in India, 39–43, 41–42, 207n24; orangutans in Borneo, 46–47; during youth, 16–17, 19

—influence: of Adventism on, 5, 14–15, 50, 127, 142, 150, 189; of Charles E. Bessey on, 20; on Aldo Leopold, 128, 143, 147, 201, 217n27; on Rosalie Edge, 183, 187–88; on younger conservationists, 6, 128, 143, 147, 182–83, 187–88, 201; on zoos, 173–74, 201–2

—taxidermy: advocates realism in, 54; buffalo group, 64; chooses as career, 19; "Fight in the Treetops" (orangutan and gibbon), 52–54; group displays, 52–54; —, promotes, 52–53, 55, 64; "Mungo" (elephant), 55; "Ole Boss" (crocodile), 26–27, 39; as taxidermist, 19, 21–22, 23, 26, 56, 57–59, 66–67, 71–72, 187

INDEX

Hornaday, William Temple (*cont.*)
—testimony: before Beck committee, 194; on behalf of National Zoological Park, 68; before House Ways and Means Committee on plumage ban, 128, 131–32; before Senate on fur seals, 108; before Senate Finance Committee on plumage ban, 133
—views/beliefs: on alcohol, 27, 37, 43, 45, 54, 180; Americanism during World War I of, 151–52; on animal intelligence, 162–63; on animal performances, 82; on Anti-Catholicism, 16, 50; on bag limits, 163–64, 171, 172, 184–85; on civil rights and crime, 193–94; conservation ideology or philosophy of, 5, 51, 66, 84–85, 127, 163, 188–89, 190, 199–201; on Germany and Germans as evil, 142, 151–52; on hunting and purpose, 17, 51, 61, 95, 103; on Indians and Hinduism, 36, 37–38; Malthusianism of, 5, 163; on migratory bird population numbers, 163, 168; on Native Americans, 8, 62–63, 65–66, 84, 103; opinion of British Imperialism, 36, 50; politics of, 149, 191–92; primitivism of, 72–73, 98; on race and ethnicity, 7–8, 27, 34, 36, 84, 91, 103–4, 158, 193–94; on scientists, 6, 20, 71, 88, 124; on seal trampling theory, 123–24; on shotgun regulations, 116–71, 122; on zoogoers, 78, 104
—works: *The American Natural History*, 20, 71, 87–91, 98, 107, 137; *Awake America!*, 151; *A Call to Action*, 190; *Camp-Fires in the Canadian Rockies*, 98–99, 103, 156; *Camp-Fires on Desert and Lava*, 103–4, 156; "Canoe and Rifle on the Orinoco," 59; *The Destruction of Our Birds and Mammals*, 83–85, 127; "Eighty Fascinating Years," 11, 57, 64, 69, 85, 148, 176, 192–94; "The Extermination of the American Bison," 65–66, 85, 116, 127; "How to Observe Quadrupeds," 71; "Hunting Ethics," 147; "A Last Call for Game Salvage," 199–200; *The Lying Lure of Bolshevism*, 152; *The Man Who Became a Savage*, 72; *Minds and Manners of Wild Animals*, 162–63; *Old Fashioned Verses*, 153; *Our Vanishing Wildlife*, 8, 127–28, 130, 132, 143, 182, 187; *Plain Truth*, 180, 190, 196; *Popular Official Guide to the New York Zoological Park*, 82, 174; "Robbed," 143; royalties from, 59, 115; "A Slump in American Morale," 177, 193; *A Square Deal for the Fur Seal*, 109; *Taxidermy and Zoological Collecting* (with Holland), 72; *Thirty Years War for Wild Life*, 85, 110, 114, 125, 136, 144, 147, 157, 168, 174, 186, 188–89, 192; *Two Years in the Jungle*, 33, 36, 43, 45, 47, 50, 59–60, 103, 174, 187, 199; —, attacked by George Pratt, 187; —, attacked in Senate by James Reed, 138–39; *Wild Animal Interviews*, 175; *Wild Life Call*, 112; *Wild Life Conservation in Theory and Practice* (with Walcott), 139–40, 172
Hornaday family history, 9–10, 204n1
Horowitz, Helen, 69, 76
Houston, David, 130, 155, 163
"How to Observe Quadrupeds" (WTH), 71
Hunt, George, 144
hunting and its regulation, 3–4; bag limits, 159–60, 163–67, 171, 176–77, 184–85, 191; Bayne non-sale of game bill (New York), 110–14; Copeland-Merritt bill, 171; Duck Stamp Tax act, 195, 200; McNary-Haugen bill, 184–86; Migratory Bird Treaty, 145–46; New-Anthony bill, 161, 167, 170–71; Norbeck bill,

INDEX

179–81, 195, 200; photography as substitution for hunting, 89, 99; —, "A Midnight Reflection" (Shiras photograph), 89; plumage, 2, 130–31; public shooting grounds, 161, 165, 169–70, 173, 179, 181, 182, 200; Underwood-Simmons Tariff Act (plumage ban), 131–36; Weeks-McLean bill/Act, 125–26, 128–30, 133, 136, 137–38, 145–46, 161, 171
"Hunting Ethics" (WTH), 147
Hyde, Anthony, 184, 191

Insull, Samuel, 165–66
Izaak Walton League, 165

Jackson, Chester, 25, 27–30, 33, 40, 42, 44, 45, 48, 49, 95
Jardin d'Acclimation, 79
Jardine, William, 176
Johnson, Robert Underwood, 3
Jones, Charles "Buffalo," 112
Jordan, David Starr, 123–25, 143, 149
Journal of Wildlife Management, 201
Junior Naval Reserve (U.S.), 148

Knapp, George, 191

Lacey, John F., 91–94, 106, 139–40
LaFarge and Heins, 80
LaGuardia, Fiorello, 171–72
Langely, Samuel P., 68–71
last buffalo hunt, 61–64, 201
"Last Call for Game Salvage, A" (WTH), 199–200
Lawyer, George, 122
League of Nations, 152
Lee, Hedges, and Company, 44–45
Lembkey, Walter, 110
Leonard, H. S., 116, 119
Leopold, Aldo, 128, 143, 147, 150, 194, 201
London, Jack, 2
Lonesome George (the last Pinta tortoise), 202
Long, William, 88, 90–91
Loring, J. Alden, 86, 93

Lucas, Frederic, 23, 53, 55, 74, 120
Lusitania, sinking of, 142
Lying Lure of Bolshevism, The (WTH), 152

MacLeod, David, 157
Madden, Dave, 56
Madsen, David, 169
Manual of Zoology (Tenney), 20
Man Who Became a Savage, The (WTH), 72
Martha (the last passenger pigeon), 1, 202
McGuire, Harry, 184–85
McIlhenny, Edward, 134
McLean, George, 135, 139, 149
McLean, Marshall, 160, 163
McNary, Charles, 184
Meine, Curt, 128, 143
Meloy, Andrew, 117
Mencken, H. L., 152
Merkel, Herman, 80
Merritt, Schuyler, 171
Migratory Bird Advisory Board, 6, 129–30, 163, 166–68, 172, 176–77, 184, 191, 195
Migratory Bird Treaty, 145–46, 161
Miller, David, 10, 11, 12, 13, 14, 18, 19, 54, 58, 101, 146, 149–51, 153
Miller, Minos, 10, 12, 13, 18
Miller, W. DeWitt, 182
Miller, William, 199
Minds and Manners of Wild Animals (WTH), 162–63
Mitchell, H. Ray, 86, 158, 195
Mondell, Frank, 125
Muir, John, 2, 3, 5, 107
Murphy, Robert Cushman, 186

Nagel, Charles, 110, 136; threatens WTH, 109
National Association of Audubon Societies, 2, 113, 126, 135, 147–48, 162, 165, 169, 182, 186; fear of WTH, 183; subscription scandal, 117–21, 148, 160, 183

National Museum (U.S.), 26, 53, 57–58, 60, 62, 66
National Wildlife Federation, 200
National Zoological Park (U.S.), 66–70
Nature Fakers, 88, 90–91, 162
Nelson, Edwin, 145, 154–56, 160–61, 166–68, 176–77
New, Harry, 161
New York Fish, Forest, and Game League, 121
New York Forest, Fish, and Game Commission, 120
New York Times, 79, 84, 94, 106, 168, 177, 180, 194; reprimands WTH, 134
New York Zoological Park (Bronx Zoo), 4, 74, 75, 80–81, 96, 104, 126, 137, 150–51, 158, 162, 173, 198; South Bronx Park, 79–80
New York Zoological Society, 4, 74, 76–79, 101, 104, 114–15, 133, 135, 140, 166, 170, 172, 174, 194, 197; as conservation organization, 83, 85
Norbeck, Peter, 179–80
North American Fish and Game Protective Association, 100

Old Fashioned Verses (WTH), 153
Oldys, Henry, 131, 134–36
Osborn, Fairfield, 198
Osborn, Henry Fairfield, 74–78, 80–83, 86, 95, 98, 101, 109, 114, 122–23, 126, 130, 134, 160, 172, 174, 192; background, 75–76; embarrassed by WTH, 102; hires WTH, 74; orders WTH to abandon attacks on Biological Survey, 166; on race, 76; restrains WTH, 166; suggests WTH adopt Hagenbeck reforms, 82; supports WTH, 112
Oskaloosa College, 19
Ota Benga, 96–98
Our Vanishing Wildlife (WTH), 8, 127–28, 130, 132, 143, 182, 187

Pablo, Michael, 105
Palmer, Theodore, 84, 112, 130, 145–46; angers WTH over Weeks-McLean Act, 130
Passing of the Great Race, The (Grant), 75–76
Peale, Charles Willson, 53
Pearson, T. Gilbert, 104, 117–20, 129, 131–36, 137, 147–48, 165, 169–70, 181, 183, 186
People Home Journal, 156–57
Permanent Wildlife Protection Fund, 114–15, 143, 146, 153, 156–57, 176–77, 184, 188, 191, 197; freedom from New York Zoological Society, 114–15, 176
Philips, John C., 66
Phillips, John M., 95–96, 99, 100, 102, 118, 158
Pinchot, Gifford, 3, 5, 93, 107, 140
Plain Truth (WTH), 180, 190, 196
Pope, Gustavus, 170
Popular Official Guide to the New York Zoological Park, (WTH), 82, 174
Punke, Michael, 62

Quinn, Davis, 182

race and conservation, 76–77. *See also under* Grant, Madison; Hornaday, William Temple; Roosevelt, Theodore
Recreation magazine, 85, 99
Red Cross, 150
Red Scare, 152–53
Reed, James, 138–39
Riggs, Thomas, 159–60
"Robbed" (WTH), 143
Roosevelt, Franklin D., 191–92, 195–96, 199
Roosevelt, Theodore, 22, 48, 61, 76–77, 88, 91–93, 102, 113, 127, 131–32, 139–41, 146, 149, 151, 154, 156, 190; association of hunting and national character by, 77; and buffalo preserves, 92, 105–6; meets WTH, 64; racial views and conservation, 7; reviews *Our Vanishing Wildlife*, 128;

INDEX

role in preserving the buffalo, 106; supports game sanctuary plan, 140
Root, Elihu, 145
Ross, Captain, 35, 36
Rothermel Committee, 110, 122–23, 125
Round the World (Carnegie), 48
"rubbish war," 104

Sage, Olivia, 119, 127
Sand County Almanac, A (Leopold), 201
Santa Fe Game Protection Association, 143
Science, 55, 60, 72, 90, 139, 163
Seton, Ernest Thompson, 2
Seymour, Edmund, 110, 149, 161, 174, 179–80, 185, 192, 195
Sheldon, Charles, 154–55
Shields, George O., 85, 89, 99, 116, 122, 130, 199
Shipley, Donald, 176
Sierra Club, 2, 3, 127
Silent Spring (Carson), 6
Silz, August, 2, 111, 188
Sitting Bull, 63
"Slump in American Morale, A" (WTH), 177, 193
Smith, Hoke, 133
Smithsonian Institution, 63, 69
Smoot, Reed, 146
Society of American Taxidermists, 54–57
Sonoran Desert, 101–3, 106
Spry, William, 144
Square Deal for the Fur Seal, A (WTH), 109
Stott, Jeffery, 81
Studer, Jacob, 56
Sulzer, Charles, 154
Sutherland, Daniel, 160

Taft, William Howard, 129
taxidermy: displays (*see under* Hornaday, William Temple [WTH]: taxidermy); effect of Morrill Act on, 19

Taxidermy and Zoological Collecting (WTH and Holland), 72
Theobald, Albert, 40
Thirty Years War for Wild Life (WTH), 85, 110, 114, 125, 136, 144, 147, 157, 168, 174, 186, 188–89, 192
Townsend, Charles, 122–24
Triangle Shirtwaist fire, 112
Trowbridge, Lawrence, 111
Two Years in the Jungle (WTH), 33, 36, 43, 45, 47, 50, 59–60, 103, 174, 187, 199; attacked by George Pratt, 187; attacked in Senate by James Reed, 138–39

Underwood, Oscar, 129, 131
Union Land Exchange, 71–72, 75
University of Iowa at Ames, 19; commemorates WTH, 22, 175
Upper Mississippi River Refuge Act, 165

Van Doren, Carl, 162
Van Name, Willard, 182, 184, 198, 201
Varner, Allen, 12, 24, 72
Verner, Samuel, 96–98

Walcott, Frederic, 172, 190, 198: claims WTH was forced to retire, 173; Committee, 190–91
Ward, Henry A., 23, 24, 25, 26, 27, 31, 32–35, 36, 38, 42, 45, 49, 52, 56–58, 60, 71, 78; background, 22; complains of WTH's writing, 46; hires WTH, 22; Natural History Establishment, 22, 23, 39, 49, 54, 82; threatens to recall WTH from Asia collecting expedition, 40; unsympathetic to WTH's complaints, 39
Warner, Charles Dudley, 131
Warren, Louis, 114
Weeks, John, 135
Wenrich, Wright, 155
Whipple, James, 120
Wild Animal Interviews (WTH), 175

Wild Life Call (WTH), 112
Wild Life Conservation in Theory and Practice (WTH and Walcott), 139–40, 172
"Wildlife in American Culture" (Leopold), 201
Williams, A. Bryan, 99–101
Williams, Gerald, 94
Wilson, Edith Axton, 135
Wilson, Woodrow, 131, 135, 149
Winchester Repeating Arms Company, 116–20
Women's Christian Temperance Union, 5, 16, 120
World War I, 141–42, 148–52, 193; Hoover Moratorium and war debts, 189

Wright, Mabel Osgood, 2, 118, 120

Yale University, 137, 139–40
Yard, Robert Sterling, 170
Yellowstone National Park, 66, 100, 168
Yerkes, Robert M., 163
Yosemite National Park, 3, 107

zoo(s): connection with imperialism, 78; public patronage, 4, 173. *See also* National Zoological Park (U.S.); New York Zoological Park (Bronx Zoo)